Animal Companions

ための総合英語

Susan Williams
×
浅井みどり
Midori Asai

NAN'UN-DO

Animal Companions
動物専門職のための総合英語

音声ファイル
無料 DL
のご案内

このテキストの音声を無料で視聴（ストリーミング）・ダウンロードできます。自習用音声としてご活用ください。
以下のサイトにアクセスしてテキスト番号で検索してください。

https://nanun-do.com　テキスト番号［ **512083** ］

※ 無線 LAN（WiFi）に接続してのご利用を推奨いたします。

※ 音声ダウンロードは Zip ファイルでの提供になります。
お使いの機器によっては別途ソフトウェア（アプリケーション）の導入が必要となります。

※ Animal Comopanions 音声ダウンロードページは
以下の QR コードからもご利用になれます。

はじめに

本書は動物専門職を目指す人たちのための総合英語テキストです。動物病院といった現場で使われる頻度の高い専門用語や表現を重点的に学びながら、スピーキング、リスニング、ライティング、リーディングの4技能を効率よく伸ばしていけるようになっています。また､各ユニットの中心となるダイアローグは獣医師、動物看護師、オーナーの3人が織り成す会話になっていますが、そのテーマは初診受付、電話予約、犬の登録の手続き、予防注射（狂犬病、混合ワクチン、フィラリア症予防薬）、マイクロチップ（AIPOへの登録方法も含む）、去勢と避妊、症状の尋ね方、投薬の説明、会計、延命拒否の同意書、ペット葬儀社との打ち合わせ、里親、グルーミング、ドッグトレーナーとのしつけの相談、など多岐に渡っていています。Unitの順番どおりに進んでいかなくても、必要なトピックをピックアップして勉強することも可能になっています。

各ユニットの構成は次のようになっています。

I. Word Builder

Dialogueに出てくる重要な単語や表現を取り上げています。

II. Before Reading and After Reading

Dialogueの内容に関する設問です。先ずはBefore Readingとして、Dialogueに入る前に設問を読んでおきます。そうすることで、内容を推測でき、また、リスニングのヒントにもなります。Dialogueをひと通り学習し終えたら、After Readingとして設問に答えてください。内容を把握できているかを確認することができます。

III. Dialogue

リスニングで穴埋めし、内容を把握したら、ペア・グループで各パートの会話練習をしてください。

IV. Try This

ユニットのテーマに関する表現の英作練習をします。単語の並べ替え問題になっています。

V. Useful Expressions / How to pronounce Veterinary Terminology / Practical Writing etc.

ユニットのテーマによってタイトルが変わっています。主に動物病院で必要となるビジネス英語表現や専門用語に関する解説と練習問題になっています。

VI. Dr. Hashimoto's Column

ユニットのテーマに関するリーディングになっています。

VII. Mini Conversation

Dialogueと合わせて練習することで、ユニットのテーマについての会話力と語彙力を増やし、強化することができます。

本書の出版にあたってご協力をお願いいたしました際に、快くお引き受けくださったヤマザキ学園理事長の山崎薫先生に深謝申し上げます。専門的な内容についてご指導いただくとともに、多大なご助言を賜りましたヤマザキ動物専門学校の講師であり獣医師の小山美弥先生に、この場を借りて厚く御礼申し上げます。また、さきがおか動物病院の院長 田村新先生、獣医師の野口ゆづる先生、難波あずさ先生、そして、ヤマザキ動物専門学校の橋本直子先生からも貴重なアドバイスをいただき、大変感謝いたしております。最後に、本書の構想から出版に至るまで大変世話になりました南雲堂の加藤敦氏に心よりお礼申し上げます。

CONTENTS

Introductory Unit
Meet the People in This Book

Mary Thomas

Mary Thomas lives in Tokyo, Japan, with her husband, Paul. Mary is an English teacher and Paul is a chef. They recently adopted a new male puppy. His name is Benny, and he is a Welsh corgi. Benny is their first dog. Mary and Paul like dogs very much. They used to live in a small apartment where pets weren't allowed. But now they live in a house with its own garden, so they decided to get a corgi.

Hashimoto Ryo

Hashimoto Ryo is a veterinarian. He works at the Hashimoto Animal Clinic in Tokyo. His specialty is small animals. Most of his patients are cats and dogs, although sometimes he also treats pets such as rabbits and hamsters. Another vet and four veterinary technicians also work at the clinic. Ryo works very long hours, but he loves his job. He has wanted to be a vet since he was a little boy.

Minami Ayaka

Minami Ayaka is a veterinary technician. She is an important part of the team at the Hashimoto Animal Clinic. Ayaka comes from Hokkaido, but now she lives in Tokyo. Her specialty is large-animal husbandry. She started to work with pets when she came to live in Tokyo. She still loves horses and cows, but now she has a pet cat named Sheba, a female ginger tabby.

Naganuma Yuta

Naganuma Yuta is a student at a vocational school for dog and cat groomers. He enjoys working with both cats and dogs. He also enjoys going to dog shows with his father and mother. The Naganumas have a dog grooming salon. Yuta's ambition is to start a mobile pet grooming service, where he will visit pet owners' homes. On the weekends, he does volunteer work helping out at a local welfare center for stray dogs.

Sato Haruka

Sato Haruka has been a dog trainer for ten years. She specializes in obedience training for pet dogs, which she calls "companion animals." She lives in a large house, because she also runs a boarding kennel for dogs to stay at when their owners are on holiday or in hospital. Haruka can board up to ten dogs at a time, and she is very proud of the warm, comfortable accommodation and the clean, secure environment she offers her "guests."

Unit 1
Benny's First Visit to the Vet

I. Word Builder

1～15の語句に適切な意味をa～oの中から選んで下線に書きなさい。

1. ______ first visit	2. ______ vet	3. ______ puppy
4. ______ checkup	5. ______ take a seat	6. ______ fill out
7. ______ registration form	8. ______ consultation room	9. ______ vaccination
10. ______ coat	11. ______ appointment	12. ______ stool
13. ______ worm	14. ______ result	15. ______ test

a. 子犬　b. 腰かける　c. 初診　d. 被毛　e. ワクチン接種 / 予防接種　f. 便
g. 獣医　h. 記入する　i. 結果　j. 検査　k. 診療申込書 / 問診票
l. 予約　m. 健康診断　n. 診察室　o. (体内の) 寄生虫

II. Before Reading and After Reading

Before Reading 次の設問を読み、内容を頭に入れてから次のページの本文を読みなさい。
After Reading 本文を読み終えたら、もう一度このページに戻り、各設問に文章で答えなさい。

1. Why has Mary Thomas brought her puppy in to the clinic?

2. Can Mary read Japanese well?

3. How old is Mary's puppy?

4. When did the puppy have his first vaccinations?

5. What does Dr. Hashimoto need to check today?

III. Dialogue 02

音声を聞いて空欄を埋めた後、ペアやグループになって役を交替しながら会話を練習しなさい。

Mary Thomas : Good morning.

Minami Ayaka : Good morning. How may I help you?

Dr. Hashimoto : This is my first visit. I've brought in our new (1) ________ for his first checkup.

Minami Ayaka : I see. Please take a seat and fill out a (2) ________ ________. Would you like the English one?

Mary Thomas : Yes, please. I can't read Japanese very well yet.

Minami Ayaka : Here you are. If you need any help, (3) ________ ________ ________.

Mary Thomas : Thank you, I will.

A few minutes later

Minami Ayaka : Ms. Thomas? Dr. Hashimoto will see you now. Please go into the (4) ________ ________.

Mary Thomas : Thank you.

In the consultation room

Dr. Hashimoto : Hello, Ms. Thomas. Nice to meet you. This must be Benny. He's so cute! (5) ________ ________ is he?

Mary Thomas : He's (6) ________ ________ old.

Dr. Hashimoto : Has he had any vaccinations yet?

Mary Thomas : Yes. He had his first (7) ________ three weeks ago.

Dr. Hashimoto : I see. His eyes, ears, and (8) ________ look fine. Please make an appointment for his second vaccinations next week. Today, I'll check his (9) ________ for worms.

Mary Thomas : Thank you.

Dr. Hashimoto : Don't mention it. Please wait out in the waiting room for the (10) ________ of the test.

IV. Try This

日本文を参考にしながら [　] 内の英語を並び替え、動物病院で必要な英語表現を完成させなさい。

1. 本日はいかがされましたか。（電話の場合は「ご用件は？」に相当する）

   [may / you / help / I / how]?

   .... ?

2. この病院にいらっしゃるのは初めてですか。

   [first / this / our / your / to / visit / is / clinic]?

   .... ?

3. お名前を伺ってよろしいですか。

   [your / may / have / name / I]?

   .... ?

4. 診療申込書（問診）にご記入いただけますか。

   [form / fill / you / the / could / registration / please / out]?

   .... ?

5. 英語版の方がよろしいですか。

   [English / like / form / would / the / you]?

   .... ?

6. 書き終わったら、こちらに持って来ていただけますか。

   [you've / bring / could / finished / back / to / it / you / me / when]?

   .... ?

7. お名前が呼ばれるまで、おかけになってお待ちください。

   Please [until / take / name / a / wait / your / and / called / seat / is].

   Please ....

would like

would like には want と同じように「…が欲しい／…したい」という意味がありますが、want よりも控えめで丁寧に自分の希望を伝えたり、相手の希望を尋ねたりすることができます。2 つを比較すると次のような違いがあります。

"I want a cup of coffee."　「コーヒーがほしい」
"I'd (=I would) like a cup of coffee."　「コーヒーをお願いします」
"Do you want anything to drink?"　「何か飲む？」
"Would you like anything to drink?"　「何かお飲み物をいかがですか？」

want は相手にストレートに伝わる強めの表現なので、I want …を多用すると、わがままで子供っぽい感じになり、また、Do you want …? を用いるとぶっきらぼうな印象を与えてしまうので、ビジネスの場面では I'd like … と Would you like …? を使えるように練習しておきましょう。

A　I'd (= I would) like …（自分の希望を相手に伝えるときに用いる）

I'd like ＋名詞 …　「…をいただけますか／…をお願いします」
I'd like ＋ to do（動詞の原形）…　「（できれば）…したいのですが」

ペアになり、次の会話文の下線部の語句を枠内の語句に入れ替えながら練習しなさい。

Student 1: Excuse me. I'd like an English form.
Student 2: OK. Here you are.

a ticket for the parking area（駐車券）
to ask about vaccinations（ワクチン接種について尋ねる）
a brochure about pet insurance（ペット保険のパンフレット）
to try this new dog food（新しいドッグフードを試す）

B　Would you like …?（相手の希望を伺ったり、相手に何かを勧めたりするときに用いる）

Would you like + 名詞 …?　「…をご希望ですか／…はいかがですか」
Would you like + to do（動詞の原形）…?　「…なさりたいですか／…されますか」

ペアになり、次の会話文の下線部の語句を枠内の語句に入れ替えながら練習しなさい。

Student 1: Would you like a pen?
Student 2: Yes, I would. Thank you very much. / No, thank you. I'm fine.

the English registration form（診療申込書の英語版）
a sample of a new cat food（新しいキャットフードの試供品）
to make an appointment for the next checkup（次回の健康診断の予約）
to attend the puppy training class（子犬のトレーニングクラスに参加する）

VI. Dr. Hashimoto's Column

[A] 次の Key Words の意味を（　　　）内に書き入れなさい。

internal parasite	(　　　　　　)	external parasite	(　　　　　　)
roundworm	(　　　　　　)	tapeworm	(　　　　　　)
hookworm	(　　　　　　)	whipworm	(　　　　　　)
heartworm	(　　　　　　)	louse (sg.) < lice (pl.)	(　　　　　　)
flea	(　　　　　　)	tick	(　　　　　　)
mite	(　　　　　　)	deworming	(　　　　　　)

[B] 次のコラムを読んで、設問に答えなさい。

Parasites that may affect your dog can be divided into two types, internal and external. Internal parasites live inside your dog's body. The five most common types are roundworm, tapeworm, hookworm, whipworm, and heartworm. Most puppies are born with roundworms, and de-worming can start as early as two weeks after a puppy's birth. External parasites live on the dog's skin and in the dog's coat. These include lice, fleas, ticks, and mites. Parasites can cause serious health problems, so make sure you take your dog to the vet to be checked as soon as possible. The vet will do a stool test and a blood test and give your dog some medicine for de-worming and for getting rid of external parasites.

1. 内部寄生虫を英語で挙げなさい。

----

2. 外部寄生虫を英語で挙げなさい。

VII. Mini Conversation（初診受付）

[A] ペアになり、初診受付の会話を練習しなさい。 

VN=Veterinary Nurse, O=Owner (Mary Thomas)

VN: Good morning. How may I help you?
O: Good morning. I found your clinic on the internet and have brought in my dog today.
VN: So, is this your first visit to our clinic?
O: Yes, it is.
VN: May I have your name, please?
O: I'm Mary Thomas.
VN: How do you spell your surname?
O: It's T-H-O-M-A-S.
VN: And your first name?
O: It's M-A-R-Y.
VN: Your dog's name, please.
O: It's Benny, B-E-N-N-Y.
VN: Thank you very much, Ms. Thomas. Please take a seat and fill out this registration form. Could you bring it back to me when you've finished?

[B] 名前を聞き取る練習をします。皆さんは Student のパートを読んで、先生に名前を尋ねてください。すると、先生が英語の名前を３つ答えるので、それぞれスペルを聞いて書き取りなさい。

Student: May I have your name, please?
Teacher: I'm ______________________________.
Student: How do you spell your surname?
Teacher: It's ______________________________.
Student: And your first name?
Teacher: It's ______________________________.

SURNAME	FIRST NAME
1.	
2.	
3.	

POINT

1 回で聞き取れない場合は、次のような表現を用いて聞き直しましょう。

Could you say that again? / Pardon? / Sorry?（もう一度、言っていただけますか）
Could you speak more slowly?（もっとゆっくりと言っていただけますか）

[C] Appendix for Unit 1（p.80）で Registration Form に必要な英語表現を学習しなさい。

Unit 2

Benny Gets His Second Vaccinations

I. Word Builder

1 ～ 16 の語句に適切な意味を a ～ p の中から選んで下線に書きなさい。

1. ______ combination vaccine	2. ______ rabies	3. ______ shot
4. ______ include	5. ______ distemper	6. ______ hepatitis
7. ______ parvovirus	8. ______ parainfluenza	9. ______ adenovirus type 2
10. ______ coronavirus	11. ______ heartworm	12. ______ prevention medicine
13. ______ transmit	14. ______ mosquito	15. ______ common
16. ______ life-threatening		

a. よくある / よく見られる	b. パルボウイルス	c. 狂犬病	d. 注射
e. アデノウイルス２型	f. フィラリア	g. 予防薬	h. 含む
i. パラインフルエンザ	j. 混合ワクチン	k. (病気などを)うつす/媒介する	i. 肝炎
m. コロナウイルス	n. ジステンパー	o. 命に関わる	p. 蚊

II. Before Reading and After Reading

Before Reading 次の設問を読み、内容を頭に入れてから次のページの本文を読みなさい。

After Reading 本文を読み終えたら、もう一度このページに戻り、各設問に文章で答えなさい。

1. Which vaccinations does Benny need to have?

2. How many vaccines are there in the combination shot?

3. Why can't Benny have both shots at one time?

4. What other vaccines does Benny need?

5. How is heartworm transmitted?

III. Dialogue

音声を聞いて空欄を埋めた後、ペアやグループになって役を交替しながら会話を練習しなさい。

Minami Ayaka : Hello, Ms. Thomas. Is Benny here for his (1) ____________________?

Mary Thomas : Yes, that's right. Which vaccinations does he need to have?

Minami Ayaka : Well, he needs to have his second (2) ____________________ ____________ and a rabies shot.

Mary Thomas : What's (3) ________________ in the combination vaccine?

Minami Ayaka : There are six different vaccines in the shot. It includes distemper, hepatitis, parvovirus, parainfluenza, adenovirus type 2, and (4) ________________________.

Mary Thomas : I see. That sounds like a lot!

In the consultation room

Dr. Hashimoto : Hello, Ms.Thomas. How's Benny doing?

Mary Thomas : He is fine, thank you. He is really growing.

Dr. Hashimoto : Good. Well, let's give him his second combination vaccine today, and we'll give him his (5) ______________ shot next week.

Mary Thomas : Can't he have both shots today?

Dr. Hashimoto : No. I'm afraid he can't. It's too much at one time.

Mary Thomas : Does he need any more after that?

Dr. Hashimoto : Yes. He needs to come back in (6) _____ ____________ for the third combination shot. And then, starting in May, he needs to take heartworm (7) ________________ ______________ once a month. That lasts until November.

Mary Thomas : What's heartworm?

Dr. Hashimoto : It's a parasite (8) __________________ by mosquitoes. It's very common here.

Mary Thomas : Is it (9) ____________________?

Dr. Hashimoto : Yes, it is. It's very dangerous. I'll give him his shot now.

Minami Ayaka : (10) ______________ _________ ____________ Benny still, please?

Dr. Hashimoto : Right. That's finished. That will be all for today. Good boy, Benny!

IV. Try This

日本文を参考にしながら [　] 内の英語を並び替え、動物病院で必要な英語表現を完成させなさい。

1. この混合ワクチンには 6 種類のワクチンが入っています。

   [different / are / vaccine / in / vaccines / there / this / six / combination].

   ....

2. この混合ワクチンに含まれているのはジステンパー、肝炎、アデノウイルスです。

   [hepatitis / includes / and / combination / distemper / this / adenovirus / vaccine].

   ....

3. フィラリア予防薬を月に 1 回服用しなければなりません。

   Your dog [prevention / a / take / needs / medicine / month / heartworm / to / once].

   Your dog ....

4. それらは犬が服用しやすいチュアブル錠です。

   They're [that / to / chewable / for / take / are / dogs / easy / tablets].

   They're ....

5. フィラリアは蚊が媒介する（蚊にうつされる）寄生虫です。

   [a parasite / by / is / mosquitoes / heartworm / transmitted].

   ....

6. ワンちゃんが動かないようにおさえていただけますか。

   [still / you / please / hold / could / your dog]?

   .... ?

7. 今日はこれで終わりです。

   [be / today / all / that / for / will].

   ....

音声をよく聞き、ワクチンに含まれる感染症を正しく発音できるように練習しなさい。 

	音節ごとにゆっくり発音（2回）	ナチュラルスピード（2回）
ワクチン：	ヴァク・シィーン vac·cine	ヴァクシィーン vaccine
狂犬病：	レェイ・ビィーズ ra · bies	レェイビィーズ rabies
ジステンパー	ディス テェン ・パァ dis · tem·per	ディステェンパァ distemper
肝炎：	ヘプ ・ァ・タァイ・ティス hep·a·ti·tis	ヘプァ タァイティス hepa ti tis
パルボウイルス：	パァー (ル) ヴォゥ ヴァイ アラス par · vo·vi·rus	パァー (ル) ヴォゥ ヴァイ アラス par vo vi rus
パラインフルエンザ：	パァ ラ イン フル エン ザ pa·ra·in·flu·en·za	パァ ラインフルエン ザ parainfluenza
アデノウイルス：	アドゥ ノォゥ ヴァイ アラス ad·e·no·vi·rus	アドゥ ノォゥヴァイアラス adenovirus
コロナウイルス：	カ ロォゥ ナァ ヴァイ アラス co·ro·na·vi·rus	カ ロォゥナァ ヴァイアラス coro na vi rus
レプトスピラ症：	レェプ ・トゥ ・スパイ・ロォゥ・シィス lep·to·spi ·ro·sis	レェプトゥスパイロォゥシィス leptospi rosis
ウイルス性鼻気管炎：	ヴァイラル ラァイ ノォゥ トゥレェイ キ アイティス vi·ral rhi·no·tra·che·i·tis	ヴァイラル ラァイノォゥトゥレェイキ アイティス viral rhinotracheitis
カリシウイルス：	カリ ィ シィヴァイァラス cal·i·ci·vi·rus	カリィシィヴァイァラス cali ci vi rus
汎白血球減少症：	パァン ルゥーカァ ピィーニィア pan·leu·ko·pe·nia	パァンルゥーカァピィーニィア panleukopenia

POINT

犬のワクチン名の前には「イヌ科の」を意味する"canine (ca · nine)"（ケェイ ナイン）をつけ、猫のワクチン名の前にはすべて「ネコ科の」を意味する"feline (fe · line)"（フィー ライン）をつけます。

ex.) イヌ肝炎 → canine hepatitis　　ネコ汎白血球減少症 → feline panleukopenia

❖ オーナーさんが混合ワクチンの内容について質問されたときに正確に答えられるよう、Appendix for Unit 2 の混合ワクチンの種類の表（pp.81 - 82）と次の英文を使って練習しなさい。

Owner: What's included in the combination vaccine for dogs / cats?

Nurse: There are [数字] different vaccines in the shot. It includes [ワクチン名].

VI. Dr. Hashimoto's Advice Column

[A] 次の Key Words の意味を（　　　）内に書き入れなさい。

bite	(　　　　　　　　)	infected animal	(　　　　　　　　)
be required by law	(　　　　　　　　)	leptospirosis	(　　　　　　　　)
viral rhinotracheitis	(　　　　　　　　)	calicivirus	(　　　　　　　　)
panleukopenia	(　　　　　　　　)	pet facilities	(　　　　　　　　)
vaccination certificates	(　　　　　　　　)		

[B] 次のコラムを読んで、設問に答えなさい。

Vaccinations protect your dogs and cats from dangerous and life-threatening diseases. Rabies is one of the most dangerous viruses. It is transmitted to people through the bite of an infected animal. That's why in most countries, rabies vaccinations for dogs are required by law. Another key to good health for your dogs and cats is the combination shot. Combination vaccines for dogs that are commonly used in Japan include vaccines for distemper, hepatitis, adenovirus type 2, parvovirus, parainfluenza, and coronavirus. If your vet thinks it's necessary, you can add some vaccines for certain types of leptospirosis. Common combination shots for cats include vaccines for viral rhinotracheitis, calicivirus, and panleukopenia.

You need to take your pet's vaccination certificates with you when you go abroad with your pet, or when you take him/her to pet facilities such as animal hotels, dog runs, and training schools. Vaccinations protect not only your pet, but also protect other animals around him/her, and your family as well.

1. ネコの混合ワクチンの対象になっている感染症を挙げなさい。

....

....

2. 予防（ワクチン）接種の証明書が必要とされるのはどのような場合ですか。

....

....

VII. Mini Conversation（予防(ワクチン)接種の予診） 06

[A] 下線部1～8は予防(ワクチン)接種の予診に関係する表現です。それぞれの英文の意味を枠内のa～hの中から選び、（　　）内に記号を書き入れなさい。

VN=Veterinary Nurse, O=Owner

VN: [1] Could you fill out this vaccine screening questionnaire? (　　)

O: Sorry, but I can't read Japanese very well. Could you help me, please?

VN: Sure. Before we give your dog his vaccination, could you answer a few questions about his general health?

O: Yes, of course.

VN: Well, first of all, [2] how is his appetite? (　　)

O: It's very good.

VN: [3] How is his stool today? (　　)

O: Normal, a little soft.

VN: [4] Is he sick today? (　　)

O: No, he isn't.

VN: [5] Has he been sick in the past month? (　　)

O: No, he hasn't.

VN: [6] Has he been vaccinated in the past month? (　　)

O: No, he hasn't.

VN: [7] Has he ever had a serious reaction to a vaccine? (　　)

O: No, he hasn't.

VN: [8] Do you have any questions about today's vaccination? (　　)

O: No, I don't.

VN: Thank you very much. That's all for now. Please take a seat over there.

a. 今日の予防(ワクチン)接種について何か質問はありますか。
b. 今日、具合が悪いですか。
c. 1か月以内に病気になりましたか。
d. こちらの予防(ワクチン)接種予診票にご記入いただけますか。
e. 予防(ワクチン)接種を受けて具合が悪くなったことはありますか。
f. 食欲はありますか
g. 1か月以内に予防(ワクチン)接種を受けましたか。
h. 今日の便はどんな具合でしたか。

[B] ペアになり、予防(ワクチン)接種の予診に関する会話を練習しなさい。

Unit 3
Benny Gets Registered and Microchipped (1)

I. Word Builder

1 ～ 16 の語句に適切な意味を a ～ p の中から選んで下線に書きなさい。

1. ______ I'm afraid ...
2. ______ a bit
3. ______ procedure
4. ______ explain
5. ______ get ... microchipped
6. ______ especially
7. ______ natural disaster
8. ______ safety
9. ______ have an accident
10. ______ get lost
11. ______ be identified
12. ______ immediately
13. ______ hurt
14. ______ implant
15. ______ grain of rice
16. ______ tiny

a. 米粒　b. 少しばかり　c. 説明する　d. すぐに / 直ちに
e. 自然災害、天災　f. 痛む　g. 事故に遭う　h. 特に / とりわけ
i. 身元が確認される　j. 迷子になる　k. 安全　l. …にマイクロチップを埋め込んでもらう
m. とても小さい　n. 手続き / 手順　o. 体内に埋め込まれる器具 / インプラント
p. 申し訳ありませんが… / あいにく…

II. Before Reading and After Reading

Before Reading 次の設問を読み、内容を頭に入れてから次のページの本文を読みなさい。
After Reading 本文を読み終えたら、もう一度このページに戻り、各設問に文章で答えなさい。

1. What does Mary want to ask Ayaka about?

2. Why does Ayaka think microchipping is really necessary?

3. How small is the microchip?

4. When is Benny going to get his microchip?

5. Does it hurt?

III. Dialogue 07

音声を聞いて空欄を埋めた後、ペアやグループになって役を交替しながら会話を練習しなさい。

Minami Ayaka : Good morning, Ms. Thomas.

Mary Thomas : Good morning. We have an (1) ________________ at 10:30, but I'm afraid we're a bit late.

Minami Ayaka : Not a problem. Are you here today for Benny's rabies shot?

Mary Thomas : Yes, and I'd like to ask about the (2) ________________ for dog registration.

Minami Ayaka : OK. I'll (3) ____________ that to you after he gets his shot.

Mary Thomas : Thank you. Also I'm thinking of getting Benny microchipped.

Minami Ayaka : Good idea. Microchipping is becoming very popular among dog owners, especially after the recent (4) ____________ ____________ in Japan. In fact, microchipping may soon be required by law for some dogs.

Mary Thomas : Oh, really? I didn't know that.

Minami Ayaka : Yes, and it's really important for Benny's (5) ____________. If he has an accident or gets lost, he can be (6) ________________ immediately.

Mary Thomas : That sounds important.

Minami Ayaka : He also has to have a chip if you plan to take him with you to the U.K. If you have any other questions, you can ask the vet.

Mary Thomas : Thank you. I'll do that.

Minami Ayaka : Dr. Hashimoto is free now. Please go straight in.

In the consultation room

Dr. Hashimoto : Hello, Ms. Thomas. If you want Benny to have a (7) ______________, we can do it today right after he gets his rabies shot.

Mary Thomas : Doesn't it (8) __________?

Dr. Hashimoto : No. It feels like a regular shot. The implant is very small, about the size of a (9) __________ ________ __________. Ayaka, could you bring one to show Ms. Thomas, please?

Minami Ayaka : Sure. ***(She brings a microchip)*** Here it is, Ms. Thomas.

Mary Thomas : Oh, I see. It's really (10) __________. Okay, let's do it, Benny!

IV. Try This

日本文を参考にしながら [　] 内の英語を並び替え、動物病院で必要な英語表現を完成させなさい。

1. 今日はワンちゃんの健康診断でいらしたのですか。

   [for / you / your / here / dog's / are / today / checkup]?

   ....?

2. その件については後ほど説明いたします。

   [to / explain / later / it / I'll / you].

   ....

3. マイクロチップはペットの安全のためにとても重要なものです。

   [for / safety / a microchip / your pet's / really / is / important].

   ....

4. 他にご質問がありましたら、いつでもご連絡ください。

   If [contact / other / anytime / us / have / questions / you / please / any].

   If ....

5. マイクロチップを埋め込む手順は通常の注射と同じです。

   The procedure for [the same / implanting / that for / a regular shot / as / a microchip / is].

   The procedure for ....

6. マイクロチップは米粒くらいの大きさです。

   [of / of / rice / about / a grain / a microchip / the size / is].

   ....

V. Useful Expressions

appointment

病院やグルーミングサロンの予約は、「時間と場所を決めて人と会うための約束」に当たるため、**appointment** を用います。予約するときは **make an appointment**、予約してあるときは **have an appointment**、予約を変更するときは **change one's appointment**、予約を取り消すときは **cancel one's appointment** となります。予約に相当する英語には他にも reservation もありますが、こちらはレストラン、ホテル、劇場や乗り物などの座席といった「場所や空間をキープする」ときに用いられます。

❖ 日本文を参考にして（　　）内に適語を入れなさい。

1. 次回の予防（ワクチン）接種の予約を忘れないようにしてください。
   Remember to（　　　）（　　）（　　　　　　）for his next vaccinations.
2. 予約されていますか。
   （　　）（　　　）（　　　）an appointment?
3. 予約を変更できますか。
   （　　　）I（　　　　）my appointment?
4. 予約の取り消しをお願いします。
   （　　　）（　　　）（　　）（　　　　）our appointment.

I'm afraid (that) ...

I'm afraid (that) …には「残念ですが…」「あいにく…」「せっかくですが…」「申し訳ありませんが…」などの意味があり、相手の要望に添えないことを丁寧に伝えたり、また、相手にとって好ましくないことを「どうやら…のようです」「(あいにく)…と思います」とやんわり伝えたりするときに用います。

❖ 次の英文を日本文に書き換えなさい。

1. I'm afraid I cannot attend the meeting tomorrow.

   ………………………………………………………………

2. I'm afraid we are closed on Wednesdays.

   ………………………………………………………………

Here you are. / Here it is.

いずれも相手に何かを渡したり、提示したりするときに「はい、どうぞ」の意味で使うことができます。微妙なニュアンスの違いを言うなら、"**Here you are.**" は対象が "**you**"、つまり相手であり、相手が求めているものを「はい、どうぞ」と渡すときに使われます。一方、"**Here it is.**" は対象が "**it**"、つまり物であり、「それはここにありますよ」と物の存在を示すとき使われ、示す物が複数であれば "**Here they are.**" になります。他にも相手に物を渡すときによく使われるフレーズに "**Here you go.**" があります。これは "**Here you are.**" とほぼ同じように使われていますが、よりカジュアルな表現とされています。

VI. Dr. Hashimoto's Column

[A] 次の Key Words の意味を（　　　）内に書き入れなさい。

unique ID number	(　　　　　　)	anesthetic	(　　　　　　)
biocompatible glass capsule	(　　　　　　)		
allergic reaction	(　　　　　　)	special scanner	(　　　　　　)
radio signal	(　　　　　　)	activate	(　　　　　　)
battery	(　　　　　　)	maintenance	(　　　　　　)
last for the rest of ... life	(　　　　　　)	permanent ID	(　　　　　　)

[B] 次のコラムを読んで、設問に答えなさい。

A microchip is a very small computer chip that is programmed with a unique ID number and implanted under your pet's skin. The process is similar to that for a regular shot and takes only a few seconds. No anesthetic is required. The microchip is in a biocompatible glass capsule, which means that allergic reactions are extremely rare. Thanks to the microchip, your pet's ID number can be read with a special scanner. When a scanner is passed over the pet's skin, it emits a radio signal that activates the microchip. The microchip then sends the ID number back to the scanner. Since it is activated by the scanner, the microchip itself requires no battery. Once a microchip is implanted, it requires no maintenance. It will last for the rest of your pet's life, acting as a permanent ID that cannot be lost. Microchipping your pet is a simple and safe way to protect him/her from getting lost or stolen. Today, many countries, including Japan, require all pets entering the country to have a microchip.

1. マイクロチップの利点を挙げなさい。

2. マイクロチップ本体に電池が不要な理由を書きなさい。

VII. Mini Conversation（電話予約）

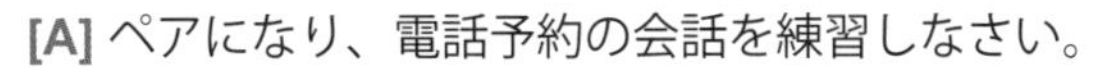
[A] ペアになり、電話予約の会話を練習しなさい。 

VN=Veterinary Nurse, O=Owner (Mary Thomas)

VN: Hello. This is the Hashimoto Pet Clinic. How may I help you?
O: I'd like to make an appointment for a physical checkup for my dog.
VN: Certainly. May I ask who's calling, please?
O: This is Mary Thomas.
VN: Hello, Ms. Thomas. When would be convenient for you?
O: April 1st would be great.
VN: That's fine. What time would be good for you?
O: How about around 11:00 in the morning?
VN: I'm afraid 11:00 is already booked. How about 11:30?
O: That sounds good.
VN: All right. Then we'll see you on April 1st at 11:30.
O: Thank you very much. See you then. Goodbye.
VN: Goodbye.

[B] 次の会話の下線部に日にち、曜日、時間を自由に入れて練習しなさい。

Student A: I'd like to make an appointment for my cat to see a vet.
Student B: Certainly. When would be convenient for you?
Student A: 日にち / 曜日 would be great.
Student B: That's fine. What time would be good for you?
Student A: How about around 時間?
Student B: Sorry, but 時間 is already booked. How about 時間?
Student A: That sounds good.
Student B: All right. Then we'll see you (on) 日にち / 曜日 at 時間.

下線部に曜日を入れるとき、例えば、来週の水曜日 next Wednesday のように next を伴うときは、最終確認（Student B の最後の台詞）のときに on をつけずに We'll see you next Wednesday at 11:30. となります。

Unit 4
Benny Gets Registered and Microchipped (2)

I. Word Builder

1～15の語句に適切な意味をa～oの中から選んで下線に書きなさい。

1. ______ register	2. ______ city office	3. ______ pet dog
4. ______ in addition to ...	5. ______ fee	6. ______ charge
7. ______ rabies vaccination tag	8. ______ dog license tag	9. ______ payment slip
10. ______ post office	11. ______ cost	12. ______ receipt
13. ______ envelope	14. ______ required form	15. ______ complete

a. 完了する　b. 飼い犬 / 愛犬　c. 封筒　d. 必要書類　e. 料金
f（費用が）かかる　g. 請求する　h. 犬鑑札　i. 役所
j. 振込用紙 / 支払伝票　k. 領収書　l. …に加えて　m. 登録する　n. 郵便局
o. 狂犬病予防注射済票

II. Before Reading and After Reading

Before Reading　次の設問を読み、内容を頭に入れてから次のページの本文を読みなさい。
After Reading　本文を読み終えたら、もう一度このページに戻り、各設問に文章で答えなさい。

1. Does Mary have to go to the city office to register her dog?

2. How much does the clinic charge for the registration and the vaccination tag?

3. What does Mary have to do first to register the microchip?

4. Does Mary have to go to the post office herself?

5. What does she have to put in the envelope and send to AIPO?

III. Dialogue

音声を聞いて空欄を埋めた後、ペアやグループになって役を交替しながら会話を練習しなさい。

After finishing the rabies shot and implanting a microchip

Dr. Hashimoto: Now you have to register Benny with the (1) ______ ______ and his microchip with *AIPO.

Minami Ayaka: Please come with me and I'll (2) ______ how to fill out the forms.

Mary Thomas: Thank you very much. Where do I have to go to register him?

Minami Ayaka: (3) ______ ______ ______ have to be registered at the city office, but you don't have to go there yourself. We can do that for you here, free of charge. You just have to (4) ______ ______ these forms. In addition to the vaccination fee, we charge ¥3,000 for the registration and ¥550 for a (5) ______ ______ ______.

Mary Thomas: All right. Thank you.

Minami Ayaka: You're welcome. You'll receive his city dog (6) ______ ______ and a rabies vaccination tag by post very soon.

Mary Thomas: I see. Now, what do I have to do for the microchip registration?

Minami Ayaka: Please fill out this registration form first. We (7) ______ ¥5,000 for doing the implant, and then you have to fill out a payment slip for the post office. That (8) ______ ¥1,000.

Mary Thomas: Do I have to go to the post office myself?

Minami Ayaka: Yes. When they give you the (9) ______, just put it in this envelope with the required form and send it to AIPO. You'll receive a (10) ______ telling you when his registration has been completed. It may take a few weeks.

＊AIPO: Animal ID Promotion Organization（動物 ID 普及推進会議）

IV. Try This

日本文を参考にしながら [　] 内の英語を並び替え、動物病院で必要な英語表現を完成させなさい。

1. 書類の記入方法をご説明します。

   [explain / fill / the forms / to / I'll / out / how].

   ....

2. 飼い犬はすべて役所に登録しなければなりません。

   [to / at / dogs / have / registered / the city office / all / be / pet].

   ....

3. 登録の手数料として 3,000 円、狂犬病予防注射済票交付の手数料として 550 円いただきます。

   We [a rabies vaccination tag / for / for / ¥550 / ¥3,000 / charge / the registration / and].

   ....

4. 犬鑑札と狂犬病予防注射済票は郵送で受け取ることになります。

   You'll [a rabies vaccination tag / receive / post / and / city dog license tag / by / your dog's].

   You'll ....

5. 領収書を必要書類と一緒にこの封筒に入れて AIPO に送ってください。

   Put [to / the required form / and / with / this envelope / in / AIPO / the receipt / it / send].

   Put ....

6. 手続きには 2、3 週間かかると思います。

   [may / a few / the procedure / weeks / take].

   ....

have to do … と need to do …

いずれも do に動詞の原形が入って「…する必要がある／…しなければならない」という意味になりますが、have to do と need to do には微妙な違いがあります。**have to do** は義務的な「しなければならない」で、話者が自ら進んで行うというより、何らかの外的な要因(規則、周りの状況、第三者など)から総合的に考えて「…せざるをえない」という場合に用います。一方、**need to do** は義務というより、内的な要因(自分の考えや意志)による必要性から「…しなければならない」という場合に用います。

✣ [　　] の話者の背景をよく読み、(　　) 内に have to もしくは need to を入れなさい。

1. I (　　　　　)(　　　) see her tomorrow. [会いたくないけど、仕事だから仕方ない]
2. I (　　　　　)(　　　) cut my hair. [髪が伸びてきたから、そろそろ切る必要がある]
3. I (　　　　　)(　　　) cut my hair. [切りたくないけど、学校の規則だから切らなければ]

fee, charge, cost

Dialogue(p.27)では支払いに関係する単語として **fee / charge / cost** という語が出てきましたが、それぞれの違いについて学んでおきましょう。**fee** と **charge** は共に「サービスに対して支払われる料金」を指していますが、医師や弁護士など専門職に対する料金には **fee** が使われることが多く、診察料は **doctor's fee** と言います。また、**charge** は動詞としても使われ「(サービスに対する料金を) 請求する」という意味になります。**cost** は「何かをするのにかかった費用の総額」を表わしています。例えば、治療費を **the cost of medical care** と言った場合は、診察料も薬代も含んだ総額の費用ということになります。動詞では「(費用が) かかる」という意味になります。料金に関する英語は他にも **price**「商品の値段・価格」、**fare**「乗り物の運賃」などがあります。

✣ 次の英語を日本語に書き換えなさい。

1. lawyer's fee (　　　　　　　　　　)
2. utility charges (　　　　　　　　　　)
3. parking charge (　　　　　　　　　　)
4. the cost of living (　　　　　　　　　　)

You're welcome.

Thank you. などお礼の言葉に対する返答で、日本語の「どういたしまして」に相当する表現です。英語学習者にはいちばん馴染み深い表現かもしれませんが、お礼に対する返答は You're welcome. だけではありません。例えば、Unit 1 の Dialogue(p.9)では Mary の "Thank you very much." に対して Dr. Hashimoto は "Don't mention it." と答えていました。その他にも **No problem. / That's OK. / No worries. / You bet. / Anytime. / (It's) My pleasure. / Not at all. / Sure.** などたくさんあります。いつも You're welcome. ばかりでなく、他の表現も使えるように練習しておきましょう。

VI. Dr. Hashimoto's Column

An annual rabies shot is required by law. When you go abroad with your pet or take him or her to pet facilities such as animal hotels, dog runs, and training schools, you need to take your pet's vaccination certificate with you. Here's a sample of the rabies vaccination certificate. How do you say each word on the certificate in English?

第　　号

① 狂犬病予防注射済証

② 所有者（管理者）　　住　所
　　　　　　　　　　　氏　名
　　　　　　　　　　　電話番号

③ 種　　類 ....　　④ 生年月日 ....

⑤ 毛　　色 ....　　⑥ 性　　別　⑦ オス ・ メス

⑧ 名　　前 ....　　⑨ 体　　格　⑩ 大 ・ 中 ・ 小

⑪ その他の特徴 ....

上記の犬に対し、狂犬病予防注射を平成　　年　　月　　日に行なったことを証明します。

⑫ 実　施　者　　住所

　　　　　　　　氏名　　　　　　　　印

① 狂犬病予防注射済証 ....

② 所有者（管理者） ....

③ 種類 ....　④ 生年月日 ....

⑤ 毛色 ....　⑥ 性別 ....

⑦ オス・メス ....　⑧ 名前 ....

⑨ 体格 ....　⑩ 大・中・小 ....

⑪ その他の特徴 ....

⑫ 実施者　住所 ....

　　　　　氏名 ....

VII. Mini Conversation（市役所での犬登録方法を説明する）

[A] ペアになり、市役所での犬登録方法を説明する会話を練習しなさい。 

VN=Veterinary Nurse, O=Owner

VN: Here is your dog's rabies vaccination certificate.
O: Thank you very much. By the way, how do I register him?
VN: Take this certificate to the city office. You may receive a dog license tag and a rabies vaccination tag there and then, or they may send them to you later. It depends on the city.
O: Could you tell me where the nearest city office is, please?
VN: Sure, I'll give you a map. Just a moment, please. *(after a while)* Here you are. I've marked the city office with a red circle.
O: That's very kind of you. Is it far (1)from the station?
VN: No, it's (2)only a five-minute walk. Don't worry, you can't miss it.
O: Thank you so much.

[B] 下線部 (1) と (2) の語句を枠内の語句に入れ替えて、会話を練習しなさい。

(1)	(2)
from here	just a ten-minute walk
from this clinic	about a fifteen-minute bus ride
from here	only two stops on the train
from my house	about ten minutes by car

Unit 5
Spaying and Neutering

I. Word Builder

1 ～ 17 の語句に適切な意味を a ～ q の中から選んで下線に書きなさい。

1. ______ spaying	2. ______ neutering	3. ______ spay
4. ______ operation	5. ______ prevent	6. ______ unwanted
7. ______ kitten	8. ______ protect	9. ______ certain
10. ______ cancer	11. ______ hormone balance	12. ______ weight gain
13. ______ diet	14. ______ advantage	15. ______ come on heat
16. ______ reduce	17. ______ discomfort	

a. 利点　b. 子猫　c. 軽減する　d. 発情する　e. 避妊手術 / 卵巣除去
f. 癌　g. 守る　h. 手術　i. 去勢手術　j. 体重増加
k. ホルモンバランス　l. 不快感　m. 防ぐ　n. ある種の　o. 食事（療法）
p. 望まれていない / 好ましくない　q. 避妊手術を施す

II. Before Reading and After Reading

Before Reading 次の設問を読み、内容を頭に入れてから次のページの本文を読みなさい。
After Reading 本文を読み終えたら、もう一度このページに戻り、各設問に文章で答えなさい。

1. What does Ayaka want to ask Dr. Hashimoto about?

2. Does Dr. Hashimoto think Sheba needs to have the operation?

3. Why does Dr. Hashimoto think it's better for Sheba's health?

4. What is one disadvantage of spaying?

5. Are there any other advantages to spaying?

III. Dialogue 11

音声を聞いて空欄を埋めた後、ペアやグループになって役を交替しながら会話を練習しなさい。

Minami Ayaka : Dr. Hashimoto, are you busy right now? Could I ask you something?

Dr. Hashimoto : Of course. Is anything wrong?

Minami Ayaka : No, everything's fine. It's about Sheba, my cat. I'd like to ask about getting her (1) ________________.

Dr. Hashimoto : Oh, I see. How old is she?

Minami Ayaka : She's six months old, and she's growing very quickly.

Dr. Hashimoto : Well, if she's ready, it'll be a good time to do it.
Why don't you (2) ________________ her in tomorrow morning? I'll take a look at her.

Minami Ayaka : Thank you. But I was wondering, does she really need to have the (3) ________________? She doesn't go outside and meet other cats.

Dr. Hashimoto : Yes, I think so. It's better for her health. If she is spayed, it will not only (4) ________________ her from having unwanted kittens, but it will also (5) ________________ her from certain cancers.

Minami Ayaka : I've heard that spaying will change her (6) ________________ balance and cause weight gain. Is that true?

Dr. Hashimoto : Yes, sometimes. That is one problem. So you will have to be careful about her diet and make sure she gets enough (7) ________________.

Minami Ayaka : Are there any other (8) ________________ to spaying?

Dr. Hashimoto : Well, if she is spayed, she will not come on heat, which will prevent the stress and (9) ________________ of that. And she won't be so noisy!

Minami Ayaka : That sounds good! Then I'll bring her in tomorrow for a checkup.
Thank you for the (10) ________________.

Dr. Hashimoto : You're welcome.

POINT

neuteringという語には「動物の生殖器を切除すること」という意味があります。よって、「去勢」と「避妊」の両方の意味を持っていることになりますが、アメリカやイギリスの動物病院のパンフレットやウェブサイトでは「去勢と避妊」を Neutering & Spaying と表記していて、neutering を「去勢」の意味で用いられているケースが多く見られます。「去勢」には他にも castration という単語もあり、「去勢済み」は castrated とも表現されます。

IV. Try This

日本文を参考にしながら [ ] 内の英語を並び替え、動物病院で必要な英語表現を完成させなさい。

1. うちの猫の避妊手術についてお訊ねしてもよろしいですか。

   Could [something / I / you / spaying / about / my cat / ask]?

   Could .... ?

2. ワンちゃんを散歩に連れて行って遊んであげたら？

   Why [with / a walk / your dog / him / take / play / and / you / for / don't]?

   Why .... ?

3. うちの猫は本当に避妊手術を受ける必要がありますか。

   [get / really / does / need / spayed / my cat / to]?

   .... ?

4. 去勢手術によって好ましくない行動が減るだけでなく、将来の疾患も予防できるでしょう。

   Neutering will [behavior / only / also / not / diseases / prevent / decrease / but / unwanted / future].

   Neutering will ....

5. 避妊手術によってホルモンバランスが変わり、体重が増えるかもしれません。

   Spaying [may / hormone balance / weight gain / her / cause / change / and].

   Spaying ....

6. ワンちゃんに毎日しっかりと運動させるようにしてください。

   You have to [sure / every / gets / enough / day / your dog / make / exercise].

   You have to ....

7. ベニーは体重増加を抑えるために、食事療法を始める（ダイエットする）必要があります。

   Benny [to / on / weight gain / has to / go / prevent / a diet].

   Benny ....

V. Useful Expressions

If+ S1 + 動詞の現在形 ～ , S2 + will + 動詞の原形 ....

"If it rains tomorrow, I will stay at home."（もし明日雨が降ったら、私は家にいるでしょう）

この文の if 節は tomorrow と未来のことを語っているのに、どうして "will rain" ではなく、現在形の "rains" になるのでしょうか。この文の if 節には次のような働きをしています。

(1)「もし雨が降ったら」→「家にいる」（「家にいる」ための条件を表している）

(2)「もし雨が降ったら」→「いる」　　（「いる」という意味の動詞 stay を修飾している）

つまり、if 節は条件を表す副詞節の働きをしています。そして、英語には時や条件を表す副詞節では未来のことでも現在形を用いるというルールがあるため、tomorrow（未来）のことでも、現在形の "rains" が用いられているのです。Dialogue（p.33）の "If she is spayed, it will protect her from certain cancers."（もしシーバが避妊手術を受けたら、ある種の癌を防ぐことができるでしょう）もこのルールに当てはまります。

✣ A ～ D にそれぞれ適切な語句を下記の語群から選び、オーナーさんに避妊、去勢、ワクチン接種、マイクロチップ、運動の利点を説明する練習をしましょう。

Student 1: What's the advantage of [A]?

Student 2: If [B] is [C], it + will + prevent (him/her) from / protect (him/her) from / reduce the chance of (him/her) + [D].

A	spaying / neutering / vaccination / microchipping / exercise
B	your dog / your cat / your pet
C	spayed / neutered / vaccinated / microchipped / given plenty of exercise
D	getting lost or stolen（失踪や盗難）/ getting life-threatening diseases（命に関わる病気）/ having hormone-related health problems（ホルモンが関係する健康問題）/ showing unwanted behaviors like aggression and urine marking（攻撃性やマーキングなど好ましくない行動）/ experiencing weight gain（体重増加）

make sure (that) S + be 動詞 / do（一般動詞）...

make sure (that) S + V は「（必ず）S が…である（be 動詞）／…する（do）ようにする」または「（忘れずに）S に…させる」いう意味で、オーナーさんにアドバイスするときに使えます。特に、Dialogue（p.33）のように "**You have to make sure (that) S + V ...**" と **have to** と一緒に用いると、アドバスの内容がとても重要なことであり、忘れずに（必ず）行なう必要があることを相手に強調して伝えることができます。

✣ 次の英文を日本文に書き換えなさい。

1. You have to make sure that your cat doesn't lick his/her wound.
（　　　　　　　　　　　　　　　　　　　　　　　　　　　　）
2. You have to make sure that your dog gets all the necessary vaccinations.
（　　　　　　　　　　　　　　　　　　　　　　　　　　　　）

VI. Dr. Hashimoto's Column

[A] コラムに出てくる病名の発音を音声に続いて練習しなさい。 12

- テス ティ キュラー キァン サー
  tes•ti•cular can•cer（精巣腫瘍）
- ペェリ エィナルグラァンド テュー マァ
  peri•anal gland tu•mor（肛門周囲腺腫）
- ビィ・ナァイン プラ・スタァティック ハァイパー・プレェイジィア
  be•nign pro•stat•ic hyper•plasia（良性前立腺肥大）
- パァイォウ・ミィー・トゥラ
  pyo•me•tra（子宮蓄膿症）
- マス タァイ ティス
  mas•ti•tis（乳腺炎）
- オゥ ヴェア リィ アン ユゥータァ リィン ヴァヂァィ ナル テュー マァ
  o•var•i•an / u•ter•ine / vag•i•nal tu•mor（卵巣／子宮／膣腫瘍）

[B] 次の Key Words の意味を（　　）内に書き入れなさい。

be destroyed	(　　　　　　)	local government	(　　　　　　)
irresponsible	(　　　　　　)	abandon	(　　　　　　)
heat period	(　　　　　　)	marking	(　　　　　　)
mounting	(　　　　　　)	aggression	(　　　　　　)
general anesthesia	(　　　　　　)	basal metabolic rate	(　　　　　　)

[C] 次のコラムを読んで、設問に答えなさい。

There are several reasons why vets recommend neutering or spaying your pets. First of all, it prevents unwanted puppies or kittens. Unfortunately, every year many dogs and cats are destroyed by the local government because irresponsible owners simply abandon them. Secondly, it protects your pets from the risk of some serious health problems. Neutering will protect your male pets from testicular cancer, benign prostatic hyperplasia, and perianal gland tumors. Spaying female animals will help reduce the risk of pyometra, mastitis, and some kinds of ovarian, uterine, and vaginal tumors. Thirdly, neutering or spaying prevents the stress and discomfort that pets experience during heat periods. In addition, neutering or spaying may help reduce your pet's unwanted behaviors like marking, mounting, and aggression that result from hormonal changes. However, we also have to keep in mind that there are some disadvantages to neutering or spaying your pet. Having any surgery and general anesthesia always carries some risk. Some animals that have had such operations may gain weight because of changes in their hormone balance and basal metabolic rate.

❖ 去勢・避妊手術のメリットとデメリットを簡単にまとめなさい。

....

....

....

VII. Mini Conversation（チワワの妊娠）

[A] 妊娠した可能性のある犬を連れてきたオーナーさんとの会話を練習しなさい。 

VN=Veterinary Nurse, O=Owner

O: Good morning. I think my (1)Chihuahua, Coco, is pregnant.
VN: Really? Let me take down some details. How many days is it since the pairing?
O: About 20 days. We mated her with another (1)Chihuahua.
VN: That's good. She should have a safe delivery. How is she feeling?
O: She's a little lethargic, and she has morning sickness.
VN: That's quite normal. (2)Her appetite and tastes in food might change, too. As the sire is also a (1)Chihuahua, she probably won't need a caesarian section.
O: When is she due?
VN: In about 40 to 45 days. Gestation in (1)Chihuahuas is usually between 60 and 65 days.
O: Will she have a difficult delivery?
VN: Don't worry! They are usually fine. I'll check her regularly, and we can talk about what you should do.

[B] 下線 (1) の語と下線 (2) の文章を、枠内の語と文章に入れ替えて会話の練習をしなさい。

(1)	(2)
Toy Poodle	She might lose her appetite and vomit more often than usual.
Pomeranian	She may become less active and gain weight.
Yorkshire Terrier	She may sleep more and have some personality changes.

pregnant 妊娠している / take down 書き留める / pairing ペアリング / mate 交尾させる / a safe delivery 安産 /lethargic 元気がない / morning sickness つわり / appetite 食欲 / tastes in food 食べ物の好み / sire 動物の雄親 / caesarian section 帝王切開 / When is she due? 予定日はいつですか / gestation 妊娠期間 / a difficult delivery 難産 / vomit 嘔吐する / personality 性格

Unit 6
Benny Has to Have an IV-Drip (1)

I. Word Builder

1 ～ 14 の語句に適切な意味を a ～ n の中から選んで下線に書きなさい。

1. ______ seem
2. ______ problem
3. ______ have an upset stomach
4. ______ diarrhea
5. ______ watery
6. ______ yellowish
7. ______ different from usual
8. ______ get dehydrated
9. ______ serious
10. ______ blood test
11. ______ weigh
12. ______ scale
13. ______ fluids
14. ______ IV-drip

a. 水分　b. 体重計　c. 体重を量る　d.（どうやら）…と思われる / …らしい
e. 点滴　f. 重篤な / 深刻な　g. 黄色っぽい　h. 脱水状態になる　i. お腹の調子が悪い
j. 下痢　k. 水っぽい　l. 問題（点）　m. いつもと違う　n. 血液検査

II. Before Reading and After Reading

Before Reading 次の設問を読み、内容を頭に入れてから次のページの本文を読みなさい。
After Reading 本文を読み終えたら、もう一度このページに戻り、各設問に文章で答えなさい。

1. What seems to be the problem with Benny?

2. What were his stools like?

3. When did it start?

4. How much does he weigh?

5. What are they going to do for him?

III. Dialogue 14

音声を聞いて空欄を埋めた後、ペアやグループになって役を交替しながら会話を練習しなさい。

Dr. Hashimoto : Hello, Ms. Thomas. What seems to be the (1) ______________?

Mary Thomas : I am not sure. Benny seems to have an (2) __________ ______________,
and he had diarrhea five or six times yesterday.

Dr. Hashimoto : I see. What were the stools like? Were they (3) ____________?

Mary Thomas : No, not watery, just very soft, and the color was a little (4) __________.

Dr. Hashimoto : When did it start?

Mary Thomas : It started about (5) ___________ ________ ________. He had diarrhea again this morning.

Dr. Hashimoto : Did he eat anything different from usual?

Mary Thomas : No, he didn't. He just ate his (6) ______________ ________ _________.

Dr. Hashimoto : It's probably nothing serious, but I'd like to do a stool test and a blood test just to make sure. Let's (7) ____________ him first, though. Ayaka, could you weigh him, please?

Minami Ayaka : All right. Come on, Benny.

Ayaka puts Benny on the scale to weigh him.

Minami Ayaka : Good boy, Benny! Let's see ... He weighs (8) ___________ ___________.

Mary Thomas : Oh, he has lost some weight.

Dr. Hashimoto : Yes, he has. I think he needs some (9) ____________ to prevent him from getting dehydrated. We'd better give him an (10) ______________.
Why don't you take a seat in the waiting room while we do it?

IV. Try This

日本文を参考にしながら [　] 内の英語を並び替え、動物病院で必要な英語表現を完成させなさい。

1. 今日はどうされましたか。

   [seems / problem / be / to / today / what / the]?

   ....?

2. 昨日からベニーの下痢と嘔吐が続いています。

   Benny [been vomiting / had / since / diarrhea / has / he's / yesterday / and].

   Benny ....

3. うちの犬が脚を引きずっていて、とても痛そうにしています。

   [to / limping / pain / and / be / my dog / a lot of / is / in / seems].

   ....

4. 何かいつもの食べ物とは違うものを食べましたか。

   [he / anything / food / from / eat / his / different / did / usual]?

   ....?

5. たいしたことはないと思いますが、検便と血液検査をしておきます。

   It's [to / but / a blood test / serious / do / I'd / a stool test / nothing / like / probably / and].

   It's ....

6. 結果が出るまで待ち合い室でおかけになってお待ちになられたらいかがですか。

   [the waiting room / why / come out / in / you / the results / don't / until / take a seat].

   ....

What seems to be the problem?

オーナーさんに「どうされましたか」と主訴を尋ねる表現です。医療現場では "What's the problem?" よりも、"**What seems to be the problem?**" の方が好まれて使われるのは、**seem** には「(どうやら)～と思われる／～のようだ」と推測の意味があるため、seem を用いて尋ねることで「(あなたの推測では)何が問題だと思いますか」となり、オーナーさんも同じく seem を用いて「(何だか) ～のようです」と自分の推測を答えやすくなるからです。Dialogue (p.39) では、Dr. Hashimoto の "What **seems** to be the problem?" という問に対して、Mary さんは "I'm not sure. Benny **seems** to have an upset stomach." (よく分かりませんが、何だかベニーはお腹の調子が悪いようです) と答えていました。とはいえ、交通事故に遭って搬送されたペットのオーナーさんに What seems to be the problem? と尋ねるのは間違いです。目で見て分かる外傷の場合は "**What happened (to your dog)?**"「どうしましたか／何があったのですか」と尋ねましょう。

❖ ペアになり、次の会話文の A ～ C に、枠内の語句を入れて練習しなさい。

Pattern 1 **Student 1:** What seems to be the problem?

Student 2: My dog seems to [A].

A
have trouble breathing (何だか呼吸が苦しそうです)
be in severe pain (何だかとても痛がっているみたいです)
have some allergies (何かのアレルギーみたいです)
be constipated (どうも便秘をしているようです)

Pattern 2 **Student C:** What seems to be the problem / the matter / wrong?

Student D: My dog's [B] seem(s) to [C].

B ear / eyes / nose / urine / stool / condition (体の調子) / pain (痛み)

C be getting worse (悪化している) / be getting cloudy (白濁してきている) / be orange in color (オレンジ色をしている) / be swollen (腫れている) / come and go (現われたり消えたりしている) / be blind (見えていない) / be darker than usual (いつもより色が濃い) / be red inside (内側が赤い) / be dry and hot (乾いて熱くなっている)

VI. Veterinary Terminology

[A] 主訴を尋ねた後、他にも症状がないか What other symptoms does he/she have?（他に何か症状がありますか）と訊いて確認していきます。次の 1 ～ 4 の表現は、オーナーさんから細かな症状を訊きだすときによく用いられる表現です。1 ～ 4 の表現の ＿＿ の中に、枠内のさまざまな病名や症状を表す語を入れ替えて、細かな症状を訊きだす練習をしなさい。（発音とさらに多くの症状については Appendix for Unit 6（pp.83 - 84）を参照）

1. Does he/she have ＿＿ ? 「…の症状はありますか？／…を患っていますか？」

   diarrhea（下痢）/ any allergies（アレルギー）/ a stomachache（腹痛）/ a fever（熱）/ any skin redness（皮膚の発赤）/ a cough（咳）/ bad breath（口臭）/ a rash（発疹）

2. Is he/she ＿＿ ? 「…していますか」

   vomiting（嘔吐している）/ limping（足を引きずっている）/ coughing（咳をしている）/ constipated（便秘している）/ in pain（痛がっている）/ wheezing（ゼーゼーしている）

3. How is his/her ＿＿ ? 「…はどんな具合ですか？」

   behavior（行動）/ appetite（食欲）/ stool（便）/ urine（尿）/ activity level（活動レベル）

4. Is/Are his/her (right/left) ＿＿ swollen (and painful)? 「…が腫れて（痛そうにして）いますか？」

   eye(s) / face / leg(s) / stomach / ear(s) / nose / knee(s) / pad(s) / gum(s) / paw(s)

[B]（　）内の日本語に合うように、細かな症状を尋ねる英文を完成させなさい。

VN=Veterinary Nurse, O=Owner

VN: What seems to be the problem?
O: My dog seems to be losing weight.
VN: How ＿＿＿＿＿＿＿＿＿＿＿＿＿＿＿＿ （食欲はどんな具合ですか？）
O: Not bad.
VN: Is ＿＿＿＿＿＿＿＿＿＿＿＿＿＿＿＿ （食べた後に吐いていますか？）
O: Yes, sometimes.
VN: Does ＿＿＿＿＿＿＿＿＿＿＿＿＿＿＿＿ （下痢していますか？）
O: No, he/she doesn't.
VN: How ＿＿＿＿＿＿＿＿＿＿＿＿＿＿＿＿ （便はどんな具合ですか？）
O: His/Her stool is dark brown and soft.
VN: Is ＿＿＿＿＿＿＿＿＿＿＿＿＿＿＿＿ （お腹が腫れていますか？）
O: No, his/her stomach isn't swollen, but he/she seems to be in pain.

VII. Mini Conversation（症状を訊く）

[A] オーナーさんから主訴と細かな症状を訊く会話を練習しなさい。 

V.S. = Veterinary Surgeon, O = Owner (Mr. Johnson), Choco = Mr. Jonson's dog.

V.S.: Hello, Mr. Johnson. What seems to be the problem with Choco today?
O: I'm not sure. She seems to be having trouble with her 1. ears.
V.S.: Is there anything else?
O: Yes, 2. her ears smell bad, and she scratches them a lot.
V.S.: When did it start?
O: 3. About two days ago.
V.S.: I see. I think 4. she may have an ear infection.
O: Is it serious?
V.S.: No, I'm sure it isn't. I'll 5. give her some ear drops. The nurse will show you how to give them to her.
O: Thank you very much.
V.S.: Don't worry. Just be careful to check and clean her 1. ears regularly.

[B] 次の①〜③の Choco の症状と獣医師の対応を日本語に書き換えた後、[A] の会話文の 1 〜 5 の下線部を①〜③の症状と獣医師の対応に変えて、会話の練習をしなさい。

① Choco's (2. Her) 1. eyes are red and she scratches them a lot. 3. It started yesterday. The doctor thinks 4. she may have an eye infection. He will 5. give her some eye drops.

② Choco's (2. Her) 1. skin is red and itchy. She has dandruff, too. 3. It started about a week ago. The doctor thinks 4. she may have dermatitis. He will 5. give her some tablets and a medical shampoo.

③ Choco's (2. Her) 1. front paw hurts. She licks it a lot, and it is bleeding. Her owner noticed it 3. a few days ago. The doctor thinks 4. she may have a splinter or a shard of glass in it. He 5. will remove it, clean the wound, and give her some ointment and an Elizabeth collar.

Unit 7
Benny Has to Have an IV-Drip (2)

I. Word Builder

1 ～ 14 の語句に適切な意味を a ～ n の中から選んで下線に書きなさい。

1. ______ be over	2. ______ come this way	3. ______ deworming
4. ______ tablet	5. ______ feel better	6. ______ medicine
7. ______ hatch	8. ______ taste	9. ______ beef
10. ______ treat	11. ______ refuse	12. ______ spit out
13. ______ mix ... with ~	14. ______ back of one's tongue	

a.（卵が）孵化する	b. ～の味がする	c. おやつ / ご褒美	d. 吐き出す	e. 駆虫
f.薬	g. こちらへどうぞ	h. …を～に混ぜる	i. 終わる	j. 拒絶する / 拒む
k.牛肉 / ビーフ	l. 具合が良くなる	m. 舌の付け根	n. 錠剤	

II. Before Reading and After Reading

Before Reading 次の設問を読み、内容を頭に入れてから次のページの本文を読みなさい。
After Reading 本文を読み終えたら、もう一度このページに戻り、各設問に文章で答えなさい。

1. What did Benny's stool test show?

2. What does Benny need to do?

3. Why does he need to take the medicine twice?

4. How can Ms. Thomas give Benny his medicine?

5. What can Ms. Thomas do if Benny refuses to take it?

III. Dialogue

音声を聞いて空欄を埋めた後、ペアやグループになって役を交替しながら会話を練習しなさい。

Minami Ayaka : Ms. Thomas, Benny's (1) ______________ is over now. Would you come this way, please?

Mary Thomas : (2) ____________ __________ ________?

Dr. Hashimoto : The blood test was fine, but the stool test showed he has worms. He needs to take two of these (3) ________________________ _______________, one today and one next week. Don't worry. He'll be (4) ________________ ______________ soon.

Mary Thomas : Thank you so much.

Dr. Hashimoto : Don't mention it. Ayaka, will you explain to Ms. Thomas how to give the tablets to Benny?

Minami Ayaka : Yes, of course. Ms. Thomas, could you wait until his (5) __________________ is ready? Then I'll show you what to do.

After a while

Minami Ayaka : Ms. Thomas, Benny's medicine is ready.

Mary Thomas : Oh, thank you.

Minami Ayaka : These are his deworming tablets. You have to give him one of them today and one (6) ______________ ________ ____________.

Mary Thomas : Why's that?

Minami Ayaka : He has worm eggs in his stool, and they will (7) ________________ in about a week, so he needs to take the medicine twice.

Mary Thomas : How do I give him the medicine?

Minami Ayaka : Well, it (8) __________________ ______________ beef, so you can just give it to him like a treat.

Mary Thomas : What if he (9) ________________ to take it or spits it out?

Minami Ayaka : You can mix it with his food, or you can open his mouth and put it on the (10) ______________ __________ __________ _________________.

Try This

日本文を参考にしながら [　] 内の英語を並び替え、動物病院で必要な英語表現を完成させなさい。

1. オペが終わったら、先生からお話があります。

   The doctor [after / over / will / you / the operation / to / is / talk].

   The doctor ....

2. レントゲン写真から、胃の中に何か異物があることが分かりました。

   [a foreign / the X-ray image / stomach / body / your dog's / showed / in].

   ....

3. 最も上手に薬を与える方法は、美味しいおやつの中に隠してしまうことです。

   The best way [a pill / a tasty treat / give / hide / to / to / in / your dog / it / is].

   The best way ....

4. フレディは関節炎のお薬を飲まなければなりません。

   Freddie [medicine / to / arthritis / take / for / needs / his].

   Freddie ....

5. 錠剤を１錠、月に１回与えてください。

   [a / to / your dog / you / month / one / once / have / tablet / give].

   ....

6. この薬は簡単に与えられます。というのも、たいていの犬はビーフ味が大好きだからです。

   This medicine [easy / the beef flavor / is / because / to / love / give / most dogs].

   This medicine ....

You have to give him one of them today and one after ten days.

処方薬（**prescribed medicine**）の投薬方法の説明は、パターンを覚えておくとそれほど難しくありません。基本は動詞です。内服薬（**oral medicine**）は "**give**"、外用薬（**external use medicine**）は "**put**"、または "**apply**" を用います。そして動詞の後に薬の形状、用量、投与回数、与えるタイミング、薬をつける（塗る・点す・貼る）部位などの情報を必要に応じて付け加えていきます。

内服薬の場合の基本形：You have to give (him/her) （必要情報）.

例）「1日2回、食後に錠剤を1錠与えてください」

You have to [give] (him/her) one tablet twice a day after meals.

- give → 内服薬の基本動詞
- one → 用量
- tablet → 薬の形状
- twice a day → 投与回数
- after meals → 与えるタイミング

外用薬の場合の基本形：You have to put/apply （必要情報）.

例1）「この軟膏を1日3回、患部につけてください」

You have to [apply] this ointment to the affected area three times a day.

- apply → 外用薬の基本動詞
- ointment → 薬の形状
- to the affected area → 薬をつける部位
- three times a day → 投与回数

例2）外用薬：「1日2回、朝と晩に目薬を点してください」

You have to [put] two drops in his eyes twice a day in the morning and in the evening.

- put → 外用薬の基本動詞
- two drops → 用量
- in his eyes → 薬を点す部位
- twice a day → 投与回数
- in the morning and in the evening → 与えるタイミング

次の処方薬の投与方法の説明でよく使われる語句の意味を書きなさい。

1. 薬の形状：

tablet ______ pill ______ capsule ______
powdered medicine ______ liquid medicine ______
syrup ______ ointment ______ eyedrops ______

2. 与える回数：

once a day (s.i.d.*) ______ twice a day (b.i.d.*) ______
three times a day (t.i.d.*) ______ a few times a day ______
once a week/month ______ every six hours ______
every other day ______

＊処方箋で使われているラテン語由来の略語（**s.i.d.** → semel in die, **b.i.d.** → bis in die, **t.i.d.** → ter in die）

3. 与えるタイミング：

in the morning and in the evening ______
after meals ______ with meals ______ before meals ______

VI. Veterinary Terminology (Medicines)

薬名を表わす 1 ～ 10 の英語に相当する日本語を枠内から選んで（　　　）に書き入れなさい。次に、その薬を説明する英文を a ～ j の中から選び、下線部に記号を書き入れなさい。

1. Analgesics （　　　　　　　　） ______
2. Anesthetics （　　　　　　　　） ______
3. Antibiotics （　　　　　　　　） ______
4. Antidiarrheal drugs （　　　　　　　　） ______
5. Antihistamines （　　　　　　　　） ______
6. Antipyretics （　　　　　　　　） ______
7. Antiseptics （　　　　　　　　） ______
8. Ear drops （　　　　　　　　） ______
9. Laxatives （　　　　　　　　） ______
10. Steroids （　　　　　　　　） ______

解熱剤	下痢止め	麻酔剤	消毒薬	抗ヒスタミン剤
鎮痛剤	抗生物質	下剤	点耳薬	ステロイド

a. are commonly used to control allergic symptoms.

b. are applied onto the surface of living tissue to reduce or prevent the risk of infection or sepsis.

c. are used to treat or prevent ear infections, especially infections of the outer ear and ear canal, such as otitis externa.

d. are used to treat and reduce pain in the body. They are also known as painkillers or pain relievers.

e. are drugs used to block the sensation of pain during a surgical procedure. They are divided into two categories: local and general.

f. are commonly used to treat a variety of medical conditions, including inflammatory, allergic, and immunologic diseases.

g. are drugs that reduce elevated body temperature in situations such as fever.

h. are used to treat or prevent bacterial infections. They work by killing bacteria or stopping them from growing.

i. are used to treat loose, watery and frequent stools.

j. produce bowel movements and relieve constipation.

VII. Mini Conversation（投薬方法の説明）

[A] ペアになり、ゴールデンレトリバーのコーディーに処方された薬の説明に関する会話を練習しなさい。

VN=Veterinary Nurse, O=Owner (Ms. Ohto)

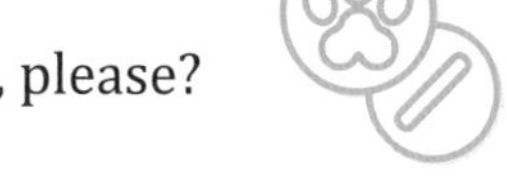

VN: Ms. Ohto, Cody's medicine is ready.

O: Oh, thank you. Could you show me how to give it to him, please?

VN: Certainly. These are (1)the tablets for his stomach.
You have to give him (2)three every morning.

O: And what about these (3)blue capsules?

VN: Those are (4)antibiotics for the infection. He needs to take (5)four, twice a day.

O: What if he refuses to take them?

VN: (6)Just mix them with his food. Golden Retrievers love to eat!

O: That's true. Good boy, Cody!

[B] 下線の語句をそれぞれ、枠内の Substitution 1 と Substitution 2 の語句に入れ替えて会話の練習をしなさい。

No.	Substitution 1	Substitution 2
(1)	the tablets for dermatitis	the drops for his eye
(2)	one in the morning and one in the evening	two drops every morning and night
(3)	brown tablets	little pink pills
(4)	steroids for the itching	antibiotics for the eye infection
(5)	one, every other day	one, twice a day
(6)	Just put them in his favorite food.	Just push them down behind his tongue and give him a treat.

Unit 8
Benny Has to Have an IV-Drip (3)

I. Word Builder

1～15の語句に適切な意味をa～oの中から選んで下線に書きなさい。

1. ______ including tax	2. ______ expensive	3. ______ cash
4. ______ pay	5. ______ credit card	6. ______ enter
7. ______ PIN	8. ______ keypad	9. ______ itemized receipt
10. ______ charge	11. ______ in detail	12. ______ altogether
13. ______ consumption tax	14. ______ insurance	15. ______ brochure

a. 総計で	b. 料金	c. 入力する	d. 領収明細書	e.（値段 / 料金が）高い
f. 詳細に	g. 保険	h. 消費税	i. 暗証番号	j. クレジットカード
k. 税込み	l. 現金	m. 支払う	n. パンフレット	o.（数字）キーパッド

II. Before Reading and After Reading

Before Reading 次の設問を読み、内容を頭に入れてから次のページの本文を読みなさい。

After Reading 本文を読み終えたら、もう一度このページに戻り、各設問に文章で答えなさい。

1. How much will Ms. Thomas be charged?

2. Will Ms. Thomas pay in cash or by credit card?

3. Can she get a receipt?

4. What are the charges for?

5. How much is the consumption tax?

III. Dialogue

音声を聞いて空欄を埋めた後、ペアやグループになって役を交替しながら会話を練習しなさい。

Mary Thomas : How much (1) ________ ________ ________?

Minami Ayaka : That'll be 6,600 yen, (2) ________ ________, please.

Mary Thomas : Oh, dear! I don't have that much cash with me.
Can I (3) ________ ________ credit card?

Minami Ayaka : Yes, of course. Please enter your (4) ________ ________ on this keypad and press the "Enter" key.

Mary Thomas : I see. Is that all right?

Minami Ayaka : Yes, fine. Now press "Enter." Here are your (5) ________ and your card. There's also an itemized receipt that explains the charges (6) ________ ________.

Mary Thomas : Could you explain the charges in English for me, please?

Minami Ayaka : (7) ________. These charges here are for Benny's tests and the IV-drip: 1,000 yen for the tests, and 4,000 yen for the IV-drip. This 1,000 yen is for his medicine, so altogether (8) ________ ________ ________ 6,000 yen.

Mary Thomas : What's this 10 percent for?

Minami Ayaka : That's the (9) ________ ________, so the total is 6,600 yen.

Mary Thomas : That's very clear. Thank you very much. By the way, I'm thinking about getting some (10) ________ ________.

Minami Ayaka : Oh, are you? I think we have a brochure on it. Here you are.

Mary Thomas : Thank you. Good-bye.

Minami Ayaka : Good-bye.

＊ PIN : personal identification number（PIN で暗証番号や個人識別番号を表わしますが、PIN number や PIN code という表現もよく使われています。）

IV. Try This

日本文を参考にしながら [　] 内の英語を並び替え、動物病院で必要な英語表現を完成させなさい。

1. (合計金額は) おいくらになりますか。

   [that / much / be / how / will]?

   .... ?

2. お支払いは現金にしますか、クレジットカードにしますか。

   [like / or / you / pay / cash / would / credit card / to / by / in]?

   .... ?

3. 申し訳ございませんが、クレジットカードはお受けできません。

   Sorry, [don't / or / credit cards / we / accept / but].

   Sorry, ...

4. 診察料とお薬代で 3,500 円になります。

   [medication / comes / the / the / 3,500 / charges / fee / to / consultation / yen / plus].

   ....

5. ペット保険はペットがもし病気になったり、事故に遭ったりした場合の治療費に適用されます。

   Pet insurance [of / pet's / accidents / covers / case / vet fees / in / illness / your / or].

   Pet insurance ....

6. 詳細については、こちらのパンフレットをご覧ください。

   [read / more / please / for / brochure / information / this].

   ....

V. How to Read Large Numbers

会計の場面に備えて、ある程度大きな桁の数字を英語で言えるように練習をしておきましょう。日本を含む漢字を使う文化圏では「一、十、百、千」の４桁を１ユニットにして、順次「万」「億」「兆」といったユニット名をつけて数えていきます。下記のように４桁毎にカンマを入れて表記すると、日本式数え方の法則がよく見えてきます。

一方、欧米の文化圏で現在一般的とされる数字の表記法では３桁を１ユニットとし、順次 "thousand, million, billion, trillion…" というユニット名をつけて数えていきます。

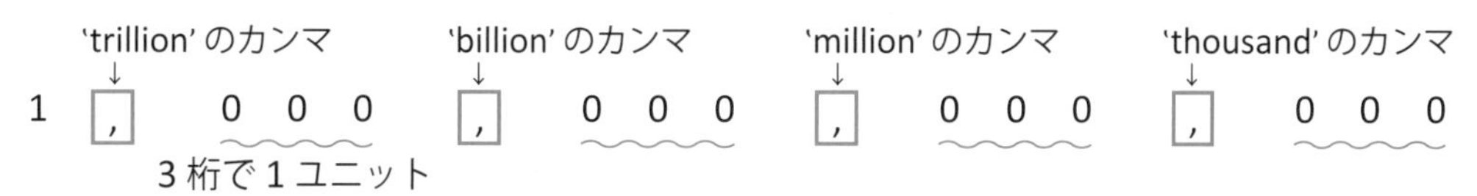

よって、英語で大きな桁の数字を読むには、先ずは基本の小単位となる１～999までの数字を練習しましょう。1,000以上の数字については、各カンマの前に数字がある場合のみ、その数字に順次 "thousand, million, billion, trillion …" というユニット名を加えて読んでいきます。

例 15,250 → fifteen [thousand] two hundred fifty

7,598,014 → seven [million] five hundred ninety-eight [thousand] fourteen

2,000,000,670 → two [billion], six hundred seventy

（million と thousand のカンマの前の３桁は 000 なのでユニット名は入れません）

次の数字を英語で書きなさい。

1～10 ______________________________

11～19 ______________________________

20 ________ 30 ________ 40 ________ 50 ________

60 ________ 70 ________ 80 ________ 90 ________

100 ______________ 205 ______________

640 ______________ 996 ______________

3,783 ______________________________

15,247 ______________________________

39,002 ______________________________

120,000 ______________________________

283,000 ______________________________

406,052 ______________________________

1,250,098 ______________________________

VI. Dr. Hashimoto's Column

[A] 次の Key Words の意味を（　　　）内に書き入れなさい。

bury	(　　　　　　)	cremate	(　　　　　　)
with dignity	(　　　　　　)	funeral	(　　　　　　)
available	(　　　　　　)	undertaker	(　　　　　　)
cemetery	(　　　　　　)	neuroscientist	(　　　　　　)
scan	(　　　　　　)	body	(　　　　　　)
grave	(　　　　　　)	recite	(　　　　　　)

[B] 次のコラムを読んで、設問に答えなさい。

Everybody wants their pet to live as long and as healthy a life as possible. Fortunately, veterinary medicine is becoming more and more advanced each year, and our pets can get the high-quality treatment and care they need and deserve. In addition, pet insurance helps cover the cost of medical treatment if our pets fall ill or are injured in an accident. Also, we need to think about what to do when our pets die. Most pet owners want to bury or cremate their pets with dignity. Recently in Japan, a variety of pet funeral services have become available. If we want to hold a funeral similar to that for human family members, all we have to do is call the pet undertakers, and they'll make all the arrangements for the ceremony.

In western countries, some owners bury their pets in a pet cemetery, while others choose to bury them at home. An American neuroscientist, Dr. Gregory Berns, who succeeded in scanning dogs' brains with *fMRI while they were awake, wrote a book about the death of his Golden Retriever, Lyra. Lyra died at the hospital, and when Dr. Berns and his family arrived home with her body, it suddenly grew dark and started to rain. Despite the rain, he and his wife spent two hours digging a grave for Lyra in their yard. They placed her body in the hole and held a small funeral with their daughters. After each family member placed a shovelful of earth in the grave, Dr. Berns recited "The Rainbow Bridge," which is a famous poem written expressly to comfort pet owners who have lost a beloved pet. We'll read the poem the next chapter.

* fMRI(functional MRI) 機能的磁気共鳴画像法

1. 最近の獣医学事情はどのようになっていますか。

....

2. ペットが亡くなったとき、オーナーさんはどのようにすることを望んでいますか。

....

....

VII. Mini Conversation（ペットフードについて）

[A] ペアになり、愛犬のドッグフードを買いにペットショップにやってきた来たトーマスさんと店員の会話を練習しなさい。

Ms. Thomas : Excuse me, I'm looking for the dog food section.

Sales Clerk : It's at the back of the store. Please come this way.

Ms. Thomas : I would like to get some (1)dry food for my Corgi.

Sales Clerk : Is he a puppy?

Ms. Thomas : (2)No. He's five years old, and he has a skin problem.

Sales Clerk : I see. Why don't you try this one? It's *specially formulated for (3)adult dogs with skin allergies.

Ms. Thomas : That looks fine. How much is it?

Sales Clerk : It's (4)¥4,600 for the 3.2-kilogram bag.

Ms. Thomas : Thank you. I'll take that one.

Sales Clerk : Thank you, ma'am. (5)That will be ¥4,600, please. The tax is already included.

Ms. Thomas : I'm sorry, but I have only a ¥10,000 note.

Sales Clerk : Don't worry, that's fine. Here's your change. (6)It's ¥5,400. Please make sure you have the correct amount.

＊specially formulated for 〜 〜のために特別調合（処方）された

[B] 下線の語句をそれぞれ、枠内の Pattern A と Pattern B の語句に入れ替えて会話の練習をしなさい。

No.	Pattern A	Pattern B
(1)	canned food for my poodle	dry food for my Labrador Retriever
(2)	Yes. He's just four months old.	No. He's very elderly, and he has painful joints.
(3)	puppies under six months	elderly dogs with arthritis
(4)	¥1,500 for the six-can pack	¥3,800 for the 4-kilogram bag
(5)	That will be ¥1,500, please.	That will be ¥3,800, please.
(6)	It's ¥8,500.	It's ¥6,200.

Unit 9
Saying Goodbye: Being Together at the End

Word Builder

1 ～ 14 の語句に適切な意味を a ～ n の中から選んで下線に書きなさい。

1. ______ elderly	2. ______ suffer from ...	3. ______ lymphoma
4. ______ be in a lot of pain	5. ______ be in a bad way	6. ______ advance
7. ______ prolong	8. ______ suffering	9. ______ heartbreaking
10. ______ decision	11. ______ option	12. ______ take care of ...
13. ______ not at all	14. ______ injection	

a. 決断	b. 苦しみ / 苦痛	c. 全く(そのようなことは)ない	d. 長引かせる	e. リンパ腫
f. 注射	g. 痛みがひどい	h. …を患う / …に苦しむ	i. 具合が悪い	j. 進行する
k. 選択	l. 悲痛な / つらい	m. …の看病をする	n. 高齢の	

II. Before Reading and After Reading

Before Reading 次の設問を読み、内容を頭に入れてから次のページの本文を読みなさい。
After Reading 本文を読み終えたら、もう一度このページに戻り、各設問に文章で答えなさい。

1. What is Henry suffering from?

2. How is he today?

3. What is Dr. Hashimoto going to do for him?

4. What does Dr. Hashimoto think is the kindest thing to do?

5. Why is Ms. Field going to come back tonight?

III. Dialogue (part I) 20

音声を聞いて空欄を埋めた後、ペアやグループになって役を交替しながら会話を練習しなさい。

Ms. Field has brought in her cat, Henry. He is very elderly and suffering from lymphoma, a common cancer in cats.

Minami Ayaka : Hello, Ms. Field. What can we do for you today? Is it Henry?

Ms. Field : Yes. He seems to be in a lot of pain, and he won't eat. He has really bad (1) __________, too.

Minami Ayaka : Please wait a moment. I'll call Dr. Hashimoto.

Dr. Hashimoto : Hello, Ms. Field. Is poor old Henry (2) _____ _____ _____ _____?

Ms. Field : Yes, I think he is.

Dr. Hashimoto : I'm afraid that the (3) __________ has advanced so far that we can't do any more for him. We'll give him an (4) __________ with something in it for the pain.

Ms. Field : How long has he got? I don't want him to (5) __________.

Dr. Hashimoto : Not very long. He is seventeen, which is very old for a cat, and we don't want to (6) __________ his suffering, either. It must be a heartbreaking decision for you to make, but the kindest option may be to (7) _____ _____ _____ _____.

Ms. Field : We think so, too, but we would all like to be there with him when it happens.

Dr. Hashimoto : Would you like to come back tonight? We'll (8) _____ _____ _____ _____ him until then.

Ms. Field : All right. I'll talk to my husband and son, and we'll come back tonight.

Dr. Hashimoto : Fine. That way, he won't suffer, and you will all get to (9) _____ _____ _____ him.

Ms. Field : When you put him to sleep, does it hurt?

Dr. Hashimoto : No, not at all. I'll give him an (10) __________, and he'll just go gently to sleep.

Ms. Field : Thank you very much.

POINT

Part VII. Mini-Conversation では、枠内の会話文を「DNA（蘇生処置拒否）」の同意書について説明する会話文に入れ替えて、練習します。

IV. Try This

日本文を参考にしながら[　]内の英語を並び替え、動物病院で必要な英語表現を完成させなさい。

1. ヘンリーはかなり痛がっているようです。

   [in / of / seems / a / Henry / pain / be / lot / to].

   .

2. これから感染症に効く点滴をヘンリーに投与します。

   [infection / we'll / for / an / give / the / something in it / Henry / with / IV-drip].

   .

3. お気の毒ですが、もうこれ以上あなたの愛猫のためにできることはありません。

   [that / I'm / do / for / any / we / afraid / your cat / more / can't].

   .

4. 愛犬を安楽死させることは、あなたが下すべく最も辛い決断の一つと言えるでしょう。

   Putting [can / the hardest / you'll / your dog / one of / have to / sleep / ever / make / be / decisions / to].

   Putting

   .

5. アメリカから戻ってくるまであなたの愛犬をしっかりとお世話します。

   [care / get / we'll / America / of / until / take / from / good / your dog / back / you].

   .

6. 最愛のペットに別れを告げるのは決して簡単なことではありません。

   [say / never / pet / goodbye / your / it's / to / to / easy / beloved].

   .

V. Practical Writing

心肺蘇生法と蘇生処置拒否の説明と同意文書の下線部に、適切な語を枠内から選んで書き入れなさい。

Owner's Information and (1) ______________(同意) for CPR or DNR

"CPR" stands for "(2) ______________________________(心肺蘇生法)." It is a procedure for a pet who has stopped breathing or whose heart has stopped beating. CPR includes (3) ______________(気管挿管), (4) ____________________(心臓マッサージ), and (5) ____________________(除細動), as well as some other invasive procedures.

"DNR" stands for "(6) ______________________________(蘇生措置拒否)." This is a decision you make to NOT allow CPR to be performed in the event that your pet stops breathing or has no heartbeat. If you choose DNR, no effort will be made to attempt to (7) ______________(蘇生させる) your pet, and he or she will die.

Animals that have survived (8) ______________________________(心肺停止) and have been successfully resuscitated are in extremely critical and unstable condition. Management of the (9) ______________________(心肺蘇生後の患者) requires vigilant monitoring and technical expertise. Neverthelss, the outcome is uncertain. There is a possiblity that your pet may not respond to CPR, or that he or she may respond initially and then suffer another arrest later. There is also a possibility that your pet may die despite CPR, and even if he or she (10) ____________(生き延びる) thanks to CPR, he or she may have some (11) ______________________________(脳障害).

I have read and understood the information above or had it explained to me to my satisfaction.

Owner's Signature: .... **Date:** ....

In the case that my pet were to suffer (12) ______________(心臓の) and/or (13) ____________(肺の) arrest (stop breathing or have no heartbeat), I request that the staff at the Animal Clinic provide (14) ______________________(救命措置) as indicated below.

□ I wish the staff of the Animal Clinic to perform CPR on my pet if he or she suffers from cardiac or respiratory arrest.

□ I DO NOT want CPR performed on my pet. I understand that if my pet stops breathing and/or his or her heart stops beating that my pet will die if CPR is not performed.

post-arrest patient / intubation / cardiopulmonary resuscitation / pulmonary / defibrillation / life-saving measures / do not resuscitate / chest compressions / brain damage / cardiopulmonary arrest / revive / consent / cardiac / survives

VI. Dr. Hashimoto's Column

[A] 次の Key Words の意味を（　　　）内に書き入れなさい。

grieve	(　　　　)	meadow	(　　　　)
vigor	(　　　　)	be maimed	(　　　　)
content	(　　　　)	except for	(　　　　)
eager	(　　　　)	quiver	(　　　　)
spot	(　　　　)	reunion	(　　　　)

[B] Key Words を参考にしながら、コラムと『虹の橋』の詩を読みなさい。

As I noted in Unit 8, there is a beautiful prose poem called "Rainbow Bridge," which tells you where your pets go after they have passed away. Those who are grieving after the loss of their beloved pets will find comfort in this lovely poem. Its original author is unknown, and you can read many versions of "Rainbow Bridge" on various websites. Here's the most popular one:

Rainbow Bridge

Just this side of heaven is a place called Rainbow Bridge.

When an animal dies that has been especially close to someone here, that pet goes to Rainbow Bridge. There are meadows and hills for all of our special friends so they can run and play together. There is plenty of food, water, and sunshine, and our friends are warm and comfortable.

All the animals who had been ill and old are restored to health and vigor. Those who were hurt or maimed are made whole and strong again, just as we remember them in our dreams of days and times gone by. The animals are happy and content, except for one small thing; they each miss someone very special to them, who had to be left behind.

They all run and play together, but the day comes when one suddenly stops and looks into the distance. His bright eyes are intent. His eager body quivers. Suddenly he begins to run from the group, flying over the green grass, his legs carrying him faster and faster.

You have been spotted, and when you and your special friend finally meet, you cling together in joyous reunion, never to be parted again. The happy kisses rain upon your face; your hands again caress the beloved head, and you look once more into the trusting eyes of your pet, so long gone from your life but never absent from your heart.

Then you cross Rainbow Bridge together ....

VII. Mini-Conversation: Dialogue (Part II)（CPR/DNRにサインする場合） 

ペットが非常に重篤な状態にあり、命の危険が懸念される場合、欧米の病院ではオーナーさんにCPR（心肺蘇生法）の説明文書とDNR（蘇生処置拒否）の同意書を渡し、サインを求める傾向にあります。また、やむを得ずペットの安楽死を希望した場合にも、同意書にサインを求めるのが一般的です。日本の動物病院でも、こうした書類を用いるようになることも考えられるために、オーナーさんにDNRの同意書について説明する会話も練習しておきましょう。

❖ Dialogue（part I）(p.57)の枠内の文章を下記の会話文に入れ替えて練習しなさい。

Dr. Hashimoto: Not very long, I'm afraid. He's 17, which is very old for a cat. There is a lot we can do to *keep him alive, but if you don't want to prolong his suffering, one option you may consider is just putting him to sleep. It's sometimes the kindest thing to do.

Ms. Field: We think so, too, but we would all like to be there with him when that happens.

Dr. Hashimoto: That's fine. Would you like to come back tonight after *surgery hours? We'll take good care of him until then, but if you are going to leave him with us, there is some paperwork for you to fill out and *sign.

Ms. Field: What's the *paperwork for?

Dr. Hashimoto: It's a *DNR agreement form.

Ms. Field: What's that?

Dr. Hashimoto: If you check this box for DNR, it means your cat will not be *resuscitated if his heart stops. He will be allowed to pass away naturally.

Ms. Field: I see. I'll sign that now, and then I'll talk to my husband and son, and we'll come back tonight.

＊keep ... alive 生き続けさせる / surgery hours 診療時間 / sign 署名する / paperwork 書類手続き / DNR agreement 蘇生処置拒否の同意書 / resuscitate 蘇生させる

For More Study

1. Appendix for Unit 9-1 (p.85) で Euthanasia Consent Form（安楽死の同意書）に必要な英語表現を学習しなさい。
2. Appendix for Unit 9-2: Mini Conversation (p.86) で、動物葬祭ディレクターに相談する会話文を練習しなさい。

Unit 10
A Forever Home

I. Word Builder

1～16の語句に適切な意味をa～pの中から選んで下線に書きなさい。

1. ______ go about	2. ______ recommend	3. ______ approve
4. ______ treat	5. ______ encourage	6. ______ object
7. ______ terrible	8. ______ breeder	9. ______ can't afford
10. ______ pedigree	11. ______ shelter	12. ______ adopt
13. ______ rescued dog	14. ______ abandon	15. ______ abuse
16. ______ do some good		

a. ひどい	b. シェルター	c. (～するだけの)余裕がない	d. 良いことをする / 役に立つ
e. 物	f. 捨てる	g. ブリーダー	h. (～するよう)促す / 仕向ける
i. 保護犬	j. 扱う	k. 血統	l. 勧める
m. 虐待する	n. 良いと思う / 良しとする	o. (物事を)を始める / ～に取り組かかる	p. 引き取る

II. Before Reading and After Reading

Before Reading 次の設問を読み、内容を頭に入れてから次のページの本文を読みなさい。

After Reading 本文を読み終えたら、もう一度このページに戻り、各設問に文章で答えなさい。

1. What does Yuka want Ayaka to do?

2. Why doesn't Ayaka approve of pet shops?

3. What kind of dogs does Yuka like?

4. Why are the dogs being looked after by the rescue organization?

5. What does Yuka think about getting a dog from an animal rescue organization?

III. Dialogue

音声を聞いて空欄を埋めた後、ペアやグループになって役を交替しながら会話を練習しなさい。

Ayaka and her old friend Yuka are in a café and talking about getting a dog.

Enomoto Yuka : I really want to get a dog, but I don't know how to go about it. I know you work at the Hashimoto Animal Clinic. Can you recommend (1) ________ ________ ________ ________?

Minami Ayaka : A pet shop? No, I can't do that. I don't approve of them. Some of them don't (2) ____________ their animals very well, and they also encourage people to think of animals as (3) ____________, not living things.

Enomoto Yuka : Really? That's terrible. I didn't know that. But if I don't go to a pet shop, how can I get a dog?

Minami Ayaka : Well, you could go to a (4) ________________. There are some very good ones, but it can be expensive. What kind of dog do you want?

Enomoto Yuka : Oh, I can't afford one with a (5) ________________. I don't mind what it looks like. I like Shiba dogs, so a Shiba mix might be good.

Minami Ayaka : In that case, it might be a good idea to adopt a rescued dog from a shelter.

Enomoto Yuka : What's a (6) ______________ __________?

Minami Ayaka : Rescued dogs are dogs that have been (7) ______________ or (8) ______________ and are being looked after by animal rescue organizations. Some of them are dogs that have lost their homes because of earthquakes or other natural disasters.

Enomoto Yuka : What a good idea! I can get a dog and have a chance (9) ________ ________ __________ __________ at the same time. Are they all older dogs?

Minami Ayaka : No. There are dogs of all types and all ages, even (10) ____________ that have been born in the shelter. Would you like to go with me to one on Saturday? We can have lunch and make a day of it.

Enomoto Yuka : Great. Let's do that!

IV. Try This

日本文を参考にしながら [ ] 内の英語を並び替え、動物専門職に必要な英語表現を完成させなさい。

1. 保護犬を引き取るにはどうすればいいのか分かりません。

   [a / don't / about / I / adopting / how / dog / go / know / to / rescued].

   ______________________________.

2. 犬を連れて行けるおすすめのカフェを紹介していただけませんか。

   [recommend / cafés / can / dog-friendly / you / any]?

   ______________________________?

3. 公共の場で犬にリードをつながないでいるのは良くないと思います。

   [the leash / I / public / approve / dogs / don't / places / having / off / of / in].

   ______________________________.

4. 地元の動物保護施設から犬か猫を引き取るのがいいかもしれません。

   [or / it / local / group / might / a dog / be / rescue / your / adopt / from / a good idea / to / a cat / animal]?

   ______________________________?

5. 保護犬はすべてワクチン接種やリハビリを受け、引き取りの準備ができています。

   [are / for / rescued / rehabilitated / adoption / all / ready / vaccinated / dogs / and].

   ______________________________.

6. 私たちは引き取ることで、困っている動物の役に立てます。

   [do / need / animals / by / in / for / we / them / good / can / some / adopting].

   ______________________________.

V. Names of Breeds: Dogs

枠内のヒントを参考にして 1 ～ 12 のアルファベットの文字を並べ替えて犬種名を完成させ、その犬種の写真を下から選び（　　）内に記号で答えなさい。（猫については Appendix for Unit 10（p.87）を参照）

Miniature Dachshund / Yorkshire Terrier / Shiba Inu / Maltese /
Golden Retriever / Miniature Schnauzer / French Bulldog / Shih Tzu/
Chihuahua / Toy Poodle / Welsh Corgi Pembroke / Pomeranian

1. yPod oTleo ________________ (　　)
2. hhuaChaui ________________ (　　)
3. SIianh bu ________________ (　　)
4. aemnnaoiPr ________________ (　　)
5. eerioeiTrrr hsYrk ________________ (　　)
6. TzShi uh ________________ (　　)
7. roFcluhBn gled ________________ (　　)
8. duhcthns riinauDeMa ________________ (　　)
9. teleasM ________________ (　　)
10. vRldteGr eiroene ________________ (　　)
11. uucneMiteni aaSrhrz ________________ (　　)
12. belrWo osekr hePmigC ________________ (　　)

a.　b. 　c. 　d.

e. 　f. 　g. 　h.

i. 　j. 　k. 　l.

VI. Dr. Hashimoto's Column

[A] 次の Key Words の意味を（　　）内に書き入れなさい。

no-kill shelter	(　　　　　　)	ten commandments	(　　　　　　)
get along with ...	(　　　　　　)	crucial to ...	(　　　　　　)
well-being	(　　　　　　)	punishment	(　　　　　　)
uncooperative	(　　　　　　)	obstinate	(　　　　　　)

[B] Key Words を参考にしながら、コラムと『犬の十戒』を読みなさい。

Fortunately, in Japan, the numbers of dogs and cats that are euthanized by the local government authorities are decreasing year by year. One of the biggest reasons for this is that public awareness of animal welfare is increasing. Furthermore, no-kill animal shelters are playing a leading role in rescuing abandoned or abused animals and in finding foster homes for them. If you want a dog, it's a good idea to adopt a dog from a shelter. Once you adopt a dog, you accept responsibility for your dog's welfare for the rest of his or her life.

The following list of rules is called "The Ten Commandments of Dog Ownership." It is written from the point of view of a dog. If you have a dog or want one, you should read the commandments and follow them.

The Ten Commandments of Dog Ownership

1. My life is likely to last ten to fifteen years. Any separation from you will be painful for me. Remember that before you invite me to join your family.
2. Give me time to understand what you want of me.
3. Place your trust in me—it's crucial to my well-being.
4. Don't be angry at me for long, and don't lock me up as a punishment. You have your work, your entertainment, and your friends. I have only you.
5. Talk to me. Even if I don't understand your words, I do understand your voice when you're speaking to me.
6. Be aware that however you treat me, I'll never forget it.
7. Before you hit me, remember that I have teeth that could easily crush the bones of your hand, but that I choose not to bite you.
8. Before you scold me for being uncooperative, obstinate, or lazy, ask yourself if something might be bothering me. Perhaps I'm not getting the right food. Or maybe I've been out in the sun too long or my heart is getting old and weak.
9. Take care of me when I get old; you, too, will grow old.
10. Go with me on difficult journeys. Never say, "I can't bear to watch it." Don't say, "Let it happen in my absence." Everything is easier for me if you are there. Remember, I love you.

VII. Mini Conversation（里親になる）

[A] Dialogue（p.63）の続きで、里親を探している犬が見つかったという会話を練習しなさい。 

Ayaka has some good news for Yuka so she decided to telephone her from the clinic.

Ayaka: Hello! Is that Yuka? It's Ayaka.

Yuka: Hi, Ayaka!

Ayaka: Hi. I have some good news for you about a dog. Are you still interested?

Yuka: Of course! Tell me about it.

Ayaka: Well, it's (1)a female dog, and she's (2)a Shiba mix.

Yuka: Oh, wow! Just what I wanted!

Ayaka: Hold on! She's (3)about one year old and medium sized, about 12 kilograms.

Yuka: What's she like?

Ayaka: She's (4)very calm and *affectionate.

Yuka: What does she look like?

Ayaka: She is (5)brown and white and short-haired.

Yuka: Sounds perfect for me. Can I come and see her?

Ayaka: Yes, of course, but adopting her may take a little time. Could you come this afternoon? Bring your name stamp. There's a lot of paperwork to get through.

Yuka: Thanks. See you later then, bye!

＊affectionate やさしい / 人懐っこい

[B] 下線 (1) ～ (5) の語句を、枠内の Pattern A と Pattern B の語句に入れ替えて会話の練習をしなさい。

No.	Pattern A	Pattern B
(1)	a male dog	a female dog
(2)	Labrador Retriever	a Chihuahua-Toy Poodle mix
(3)	a puppy and very large, about five kilograms	about six months old and very small, about one kilogram
(4)	very affectionate and playful	extremely active and playful
(5)	dark gold and long-haired	black and white with short, curly hair

Unit 11
Beautiful and Clean

I. Word Builder

1 ～ 16 の語句に適切な意味を a ～ p の中から選んで下線に書きなさい。

1. ______ shampoo	2. ______ trim	3. ______ except ...
4. ______ cater for	5. ______ breed	6. ______ in particular
7. ______ neither A nor B	8. ______ grooming salon	9. ______ service
10. ______ pick up	11. ______ bring back	12. ______ convenient
13. ______ provide	14. ______ crate	15. ______ call back
16. ______ for now		

a. 提供する / 用意する	b. 車で迎えに行く	c. 今のところ	d. サービス
e. トリミングする	f. 電話をかけなおす	g. シャンプーする	h. 引き受ける / (要求に) 応じる
i. …を除く	j. 特に	k. 便利な	l. グルーミングサロン
m. A も B も～ない	n. 送り返す	o. クレート	p. 犬種

II. Before Reading and After Reading

Before Reading 次の設問を読み、内容を頭に入れてから次のページの本文を読みなさい。

After Reading 本文を読み終えたら、もう一度このページに戻り、各設問に文章で答えなさい。

1. What does Mr. Edwards want to ask about?

2. When is the salon open?

3. Mr. Edwards can't take his dog to the salon. Why not?

4. What special service does the salon provide?

5. What details does Yuta need to know in order to provide the pick-up service?

III. Dialogue 

音声を聞いて空欄を埋めた後、ペアやグループになって役を交替しながら会話を練習しなさい。

Mr. Edwards wants to have his dog, Sandy, shampooed and trimmed so he telephones a grooming salon that he found on the Internet. The groomer, Naganuma Yuta, answers.

Naganuma Yuta : Good morning. How may I help you?

Mr. Edwards : Good morning. I'd like to ask about having my dog (1) ____________ ________ ____________. Could you give me some details about your salon and services, please?

Naganuma Yuta : Certainly, sir. We're open every day except (2) ____________, and we cater for all breeds of any size.
Is there (3) ____________ ________ ____________ you'd like to know?

Mr. Edwards : Well, yes ... There is one problem. Neither my wife nor I can drive, and your grooming salon is too far for us to walk.

Naganuma Yuta : Oh, don't worry about that, sir.
We have a (4) ____________ ____________ ____________.
We can pick up your dog in the morning at (5) ____________ ________ ____________ and bring him back to you in the evening (6) ____________ ____________ ____________.

Mr. Edwards : I see. That's wonderful. How (7) ____________________! Does he need a crate or carriage cage?

Naganuma Yuta : No, sir. We can provide the (8) ____________. If you do decide to use our pick-up service, we will need to ask for some details about your dog and as well as your (9) ____________, ____________, ________ ____________ ____________.

Mr. Edwards : That's fine. Thank you for all your help. I'm going to talk to my wife, and I'll call you back in a little while to make an (10) ____________. Goodbye for now.

Naganuma Yuta : Thank you, sir. I'll be waiting for your call. Goodbye, then.

IV. Try This

日本文を参考にしながら [　] 内の英語を並び替え、グルーミングサロンで必要な英語表現を完成させなさい。

1. うちの犬のシャンプーとトリミングをお願いしたいのですが。

   [and / have / like / shampooed / I'd / my dog / trimmed / to].

   ....

2. 火曜日と祝祭日は休業日になっております。

   [on / are / national holidays / we / and / closed / Tuesday].

   ....

3. チワワからゴールデンレトリバーまで、あらゆる大きさの犬種に対応しております。

   [from / cater for / Chihuahuas / we / breeds / size / Golden Retrievers / to / any / of / all].

   ....

4. 追加料金で送迎サービスをご利用いただけます。

   [available / drop-off / for / a / and / charge / service / pick-up / is / a slight additional].

   ....

5. 愛犬を連れていらっしゃるときには、予防注射済証を忘れずにお持ちください。

   [vaccination / to / with / bring / certificate / please / you / remember / the] when you bring your dog.

   ....

   .... when you bring your dog.

6. グルーミングサービスにはすべて肛門腺絞り、爪切り、耳掃除が含まれています。

   [nail (or claw) trimming / includes / grooming / ear cleaning / each / anal gland expression / service / and].

   ....

V. Useful Expressions

What would you like us to do today?

❖「今日はどのようにされますか」と尋ねる表現です。Owner 役の「…をお願いします」という表現と共に練習しなさい。

Pet Groomer: What would you like us to do today?

Owner: I'd like to have [A] .

A	my dog shampooed and blow-dried（シャンプーとブロー） my dog flea-tick shampooed and trimmed（ノミ取りシャンプーとカット） my dog's nails/claws clipped and his paws trimmed（爪切りと足裏の毛のカット） my dog's mats removed（毛玉取り） my dog done as usual（いつもの通り）

What style would you like?

❖「どのようなスタイルをご希望ですか」と具体的なカットの希望を尋ねる表現とカットの特徴を説明する表現を練習しなさい。

Pet Groomer: What style would you like?

Owner: What cut would you recommend for my dog?

Pet Groomer: How about [B] ? It [C] .

B	a style that is short and easy to care for（短く手入れのしやすいスタイル） a short cut for the summer（夏向けの短めのカット） a Teddy Bear Cut（テディベアカット）/ a Lion Cut（ライオンカット） a Summer Cut（サマーカット）/ a Puppy Cut（パピーカット）
C	is one of the most popular cuts for poodles（プードルに最も人気のあるカット） prevents skin rashes and other irritations（発疹などの皮膚炎を防ぐ） is suitable for all breeds（あらゆる犬種に適している） is good in hot weather, and it does not require much maintenance （暑い時期に向いていて、お手入れもあまり必要としない） is good for long-haired dogs and cats（長毛種の犬や猫に向いている） is easy to maintain and keeps your dog's coat free from mats and tangles（お手入れしやすく、毛玉やもつれを防いでくれる）

VI. How to Wash Your Dog

Key Words を参考にしながら、愛犬を洗う 1 ～ 8 の手順を日本語に直しなさい。次にオーナーさんに説明できるよう、英文を読む練習をしなさい。

1. Brush and comb your dog's coat before the bath to remove tangles and mats.

   (　　　　　　　　　　　　　　　　　　　　　　　　　　　　　　　　　　　)

2. Wet your dog's body with warm water.

   (　　　　　　　　　　　　　　　　　　　　　　　　　　　　　　　　　　　)

3. Apply a shampoo designed for dogs.

   (　　　　　　　　　　　　　　　　　　　　　　　　　　　　　　　　　　　)

4. • Lather your dog, beginning at the neck and working back to the tail.

   (　　　　　　　　　　　　　　　　　　　　　　　　　　　　　　　　　　)

   • Make sure to cover the legs, stomach, chest, tail, and the outside of the ears.

   (　　　　　　　　　　　　　　　　　　　　　　　　　　　　　　　　　　)

   • Be careful not to get shampoo in your dog's eyes, ears, nose, or mouth.

   (　　　　　　　　　　　　　　　　　　　　　　　　　　　　　　　　　　)

   • If your dog's face is dirty, wipe it with a damp cloth.

   (　　　　　　　　　　　　　　　　　　　　　　　　　　　　　　　　　　)

5. Rinse the shampoo thoroughly out of his or her coat.

   (　　　　　　　　　　　　　　　　　　　　　　　　　　　　　　　　　　　)

6. Apply a conditioner designed for dogs.

   (　　　　　　　　　　　　　　　　　　　　　　　　　　　　　　　　　　　)

7. Rinse again with warm water until all the conditioner is out of the coat.

   (　　　　　　　　　　　　　　　　　　　　　　　　　　　　　　　　　　　)

8. Dry your dog thoroughly with towels and a hair dryer.

   (　　　　　　　　　　　　　　　　　　　　　　　　　　　　　　　　　　　)

Key Words

brush ブラシをかける / comb くしでとかす / remove 取り除く / tangle もつれ、絡み / mat 毛玉 / wet ぬらす / apply つける / shampoo designed for dogs 犬用シャンプー / lather (石鹸などの) 泡を塗る、(泡立てて) 洗う / make sure 忘れずに…する / cover 覆う / wipe 拭き取る、ぬぐう / damp cloth 湿った (ぬれた) 布 / rinse すすぐ、洗い落とす / thoroughly 完全に / conditioner コンディショナー / dry 乾かす

VII. Mini Conversation（グルーミングサロンで）

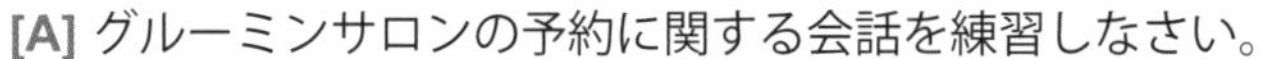
[A] グルーミンサロンの予約に関する会話を練習しなさい。

Naganuma Yuta at the salon gets a telephone call from a customer who would like to use Yuta's his pick-up service. Yuta needs to get some details from the customers.

Naganuma Yuta : Good morning. How may I help you?

Mr. Edwards : Hello. This is Mr. Edwards. I called yesterday, and I'd like to make an appointment for your pick-up service for tomorrow, please.

Naganuma Yuta : Thank you very much, sir. May I just get some details from you?

Mr. Edwards : Certainly.

Naganuma Yuta : Could you tell me your dog's name, size, and breed?

Mr. Edwards : Yes. He's called Duke, he's about thirty-five kilograms, and he's a German Shepherd.

Naganuma Yuta : All right. Great name! And now we just need to know your full name, address, and telephone number.

Mr. Edwards : It's David Edwards at 1-2-3, Fujimicho, Setagaya-ku, Tokyo. My mobile number is 090-2345-67XX.

Naganuma Yuta : Thank you very much. Just one more thing, Mr. Edwards. Are Duke's vaccinations up to date?

Mr. Edwards : Yes, they are.

Naganuma Yuta : Fine. Then we'll be at your house tomorrow morning between nine and ten o'clock to pick Duke up.

Mr. Edwards : Thank you. See you tomorrow.

[B] 下の表に自分に関する情報を英語で記入しなさい。次に上記の会話の下線部に自分の情報を入れて、会話文を読む練習をしなさい。

名　前			電話番号		
住　所					
犬の名前		犬　種		ワクチン接種	未 ・ 済
犬の体重			送迎希望日		

POINT

英語で住所を書く場合は日本の表記とは逆に、狭い場所から大きな場所に向かって書いていきます。

建物名 → 部屋番号・階 → 丁目・番地・号 → 地域名 → 市区町村 → 都道府県 → 郵便番号

ex.) 〒 162-0801 東京都新宿区山吹町１－２－３　アニマルビルディング７階
→ Animal Bldg. 7F, 1-2-3, Yamabuki-cho, Shinjuku-ku, Tokyo, 162-0801

（階数は、建物名の前に書いても構いません。例：7F Animal Building, 1-2-3 ...）

Unit 12

Sammy Has Obedience Classes

I. Word Builder

1～16 の語句に適切な意味を a～p の中から選んで下線に書きなさい。

1. ______ manage	2. ______ be worried about ...	3. ______ pull on the leash
4. ______ command	5. ______ obey	6. ______ lively
7. ______ intelligent	8. ______ trainable	9. ______ method
10. ______ concentrate on ...	11. ______ one at a time	12. ______ reward
13. ______ fantastic	14. ______ confident	15. ______ calm
16. ______ consistent		

a. … に専念する / 集中する	b. しつける / 扱う	c. 落ち着いた
d. 活発な / 元気な	e. 指示	f. しつけやすい / 訓練しやすい
g. 素晴らしい	h. 従う	i. リードを引っ張る
j. 一貫性のある	k. 方法	l. (今)… が気になっている
m. ひとつずつ	n. 自信がある	o. 利口な
p. ご褒美		

II. Before Reading and After Reading

Before Reading 次の設問を読み、内容を頭に入れてから次のページの本文を読みなさい。

After Reading 本文を読み終えたら、もう一度このページに戻り、各設問に文章で答えなさい。

1. What is Jane most worried about?

2. Which commands does Sammy understand?

3. Where will Sammy have his obedience classes?

4. How far will they be able to go in six weeks?

5. What are the three "C's"?

III. Dialogue

音声を聞いて空欄を埋めた後、ペアやグループになって役を交替しながら会話を練習しなさい。

Jane Monroe has a six-month-old Labrador Retriever puppy. At 25 kilograms he is not yet fully-grown, but he is already getting big enough to be difficult to control. She'd like to find a dog trainer to help her manage him. A friend has recommended a place called "Gentle Companions," and today Jane is going to meet one of the trainers, Ms. Sato Haruka, at the trainer's office.

Jane Monroe : Is this "Gentle Companions"? Ms. Sato?

Sato Haruka : Yes. Please come in. You (1) ____________ ________ Ms. Monroe. Please call me Haruka.

Jane Monroe : Thank you, I will. And please call me Jane. I've come about my dog, Sammy.

Sato Haruka : I see. Tell me something about him.
What are you (2) ________________ ________________ the most?

Jane Monroe : Well, he's a Lab and he's six months old, and he weighs 25 kilograms and is very strong. He (3) ____________ ________ ________ ______________ and he won't do anything I tell him. It's becoming difficult and dangerous!

Sato Haruka : Does he understand any (4) ______________________?

Jane Monroe : Yes. He understands "Sit!" and "Down!" but he doesn't usually obey me.

Sato Haruka : Does he come when you (5) ______________ ____________ ______________?

Jane Monroe : Not usually. Only when he knows it's dinner-time!

Sato Haruka : Right. I think I understand your problem. It's very common in young,lively dogs. (6) ____________ ____________ __________ try a six-week course? That should be enough. You'd be surprised at how intelligent and trainable Labs are.

Jane Monroe : Thank you. Can you tell me more about your training (7) ______________?

Sato Haruka : Yes, of course. First of all, I will come to your house for the classes. We will concentrate on simple commands at first, one at a time with a system of treats as (8) ______________.

Jane Monroe : How far can we go in six weeks?

Sato Haruka : Well, we should be able to cover the (9) ____________ __________________ such as, "Sit!," "Wait!," "Down!," "Come!," and "Fetch!" and we'll make sure he learns to walk without pulling.

Jane Monroe : That sounds (10) ________________________! And what a relief!

Sato Haruka : Yes. But please keep in mind, Jane, that the most important part of the training will be you, the owner! You yourself will have to master the three "Cs."

Jane Monroe : The three "Cs"?

Sato Haruka : Yes. For Sammy's sake, you have to be "Confident," "Calm," and "Consistent"!

IV. Try This

日本文を参考にしながら [　] 内の英語を並び替え、犬のトレーニングに必要な英語表現を完成させなさい。

1. 愛犬をコントロールするのが難しくなってきたら、いつでもご相談ください。

   Please feel free to contact us [to / is / control / dog / difficult / if / getting / your].

   Please feel free to contact us ....

2. こんにちは、スミス様でいらっしゃいますね。

   Hello, [must / Ms. / you / Smith / be].

   Hello, ...

3. 愛犬のことで今いちばん気になっていることは何ですか。

   [with / about / are / most / what / dog / you / your / the / worried]?

   .... ?

4. このトレーニングコースで、愛犬は基本的な指示に従うことを学びます。

   In this training course, [to / learn / commands / your / learn / will / basic / dog / follow].

   In this training course, ...

5. 資格を持ったドッグ・トレーナーが、プライベートレッスンをするためにお宅へ伺います。

   Our [private / certified / will / your house / dog trainer / lessons / to / come / to give].

   Our ....

6. 犬があなたに従ったら、ご褒美をあげるのを忘れないように。

   [forget / him / your / you / when / give / don't / dog / a treat / obeys / to].

   ....

VI. Essential Dog Commands

警察犬 (**police dog**)、盲導犬 (**guide dog**)、聴導犬 (**hearing dog**)、介助犬 (**service dog**)、探知犬 (**detection dog**)、さらには、感情支援動物 (**emotional support animal**) といった working animal の訓練には英語のコマンドが用いられています。英語の指示は短く分かりやすいので、誰に指示されても犬が混乱することなく理解できるからです。最近では、一般家庭でも英語のコマンドでしつけるオーナーさんが増えつつあるため、基本的な英語のコマンドを覚えておきましょう。

❖ 1 ～ 15 の英語のコマンドを日本語に直した後、声に出して練習しなさい。

1. Watch me. ____________________
2. Sit. ____________________
3. (Lie) Down. ____________________
4. Stay. ____________________
5. Stand (up). ____________________
6. Come. ____________________
7. Heel. ____________________
8. Fetch. ____________________
9. Drop it. / Release. ____________________
10. Leave it. ____________________
11. Stop. / Hold it. ____________________
12. Off. ____________________
13. Okay. ____________________
14. No. ____________________
15. Go to bed. (or other place) ____________________

VI. Common Dog Behavior Problems

1～13の英語は犬の問題行動を表しています。それぞれを表わす絵を下から選び（　）内に記号を書き入れなさい。答え合わせの後、日本語でどのように表現するかを調べ、下線部に書きなさい。

1. Barking (uncontrollably) (　　) ____________________
2. Destructive chewing (　　) ____________________
3. Digging (　　) ____________________
4. Begging for food (　　) ____________________
5. Inappropriate elimination (　　) ____________________
6. Chasing (cars and other moving things) (　　) ____________________
7. Separation anxiety (　　) ____________________
8. Jumping up (on people) (　　) ____________________
9. Eating feces / Coprophagia (　　) ____________________
10. Leash pulling (　　) ____________________
11. Biting, nipping, mouthing (　　) ____________________
12. Stealing food (　　) ____________________
13. Running away (　　) ____________________

a.

b.

c.

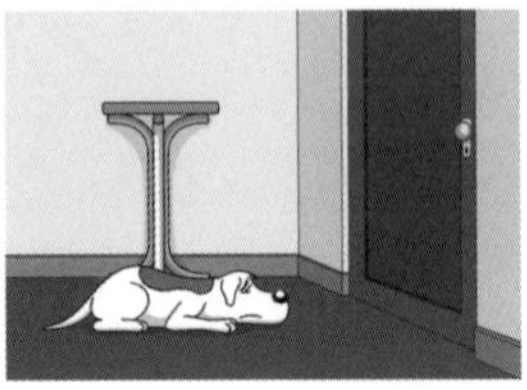

d.

e.

f.

g.

h.

i.

j.

k.

l.

m.

VII. Mini Conversation 

[A] Dialogue (p.75)に登場したトレーナーのハルカさんが、ジェーンさんの家を訪れて「お座り」とリードをつけて室内、および屋内を歩く指導をしている会話です。ペアになって練習しましょう。

Haruka : Let's try getting Sammy to sit today. Did you get some treats?
Jane : Yes, I've got some small pieces of chicken. He loves chicken. It's his favorite.
Haruka : Right. Put a small piece in your hand and close your fist over it. Then let him smell your hand. Now point the finger of your other hand and say, "Sit!"
Jane : Oh, he won't do it. He just jumps up.
Haruka : That's all right. Try again. This time move your fist over his head. As he looks up, he should sit down naturally.
Jane : Do I give him the treat then?
Haruka : Yes, and say, "Good boy, Sammy!" Then pat him and give him a little cuddle. Watch me try.
Jane : Okay. Oh! He did it. That's amazing! Good boy, Sammy! Now is it all right if I try?
Haruka : Sure. Go on, your turn.
Jane : He did it! Good boy! Oh, he did it again. He really understands now.
Haruka : Yes, he does. You have to practice it a few times every day. Now let's try walking him on the leash. Is he getting any better?
Jane : Yes, he's a little better, but he still gets very excited when he sees other dogs.
Haruka : That's normal to begin with. Let's go out and see how he behaves. Come on, Sammy! Walkies!

[B] ペアになり、トレーニングに関する短い会話を練習しなさい。

Owner: How can I teach my dog (1)to sit? I've got some treats.
Trainer: Hide the treat in your fist, and when he comes, (2)move it back over his head and say, (3)"Sit!"
Owner: Oh! He (4)sat down right away. That's amazing!
Trainer: Don't forget to praise him!

下線 (1)～(4) の語句を、枠内の Pattern A と Pattern B の語句に入れ替えて会話の練習をしなさい。

No.	Pattern A	Pattern B
(1)	to lie down	to come
(2)	move it down to the floor	move back away from him
(3)	"Down!"	"Come!"
(4)	lay down	came

Appendix for Unit 1: Registration Form

英語の診療申込書に含まれる各項目の意味を調べ、1～24 の（　　）内に日本語を書き入れなさい。

Registration Form

Owner's Name

Address

Telephone **Emergency Contact** ([1])

Email

Pet's Name

Species ([2]) □ dog □ rabbit □ cat □ ferret ([3])
□ bird □ reptile ([4]) □ other ([5])

Breed ([6]) **DOB** ([7]) **Color**

Sex ([8]) □ Male □ Female
□ Neutered Male ([9]) □ Spayed Female ([10])

What food does your pet eat?
□ Dried ([11]) □ Canned ([12]) □ Homemade

Where do you keep your pet?
□ Inside ([13]) □ Outside ([14])

◈ **Has your pet had any vaccinations in the past?**
([15])
□ Yes [□ Rabies ([16]) □ Combination Vaccine ([17]) □ Other]
□ No □ Don't know

◈ **Is your pet on heartworm medication?** ([18])
□ Yes □ No □ Don't know

◈ **Is your pet on any other medication or supplements?**
([19])
□ Yes [If so, please list here.:] ([20])
□ No □ Don't know

◈ **Is your pet allergic to anything?** ([21])
□ Yes [If so, please list here.:] □ No □ Don't know

◈ **Has your pet had any serious illnesses?** ([22])
□ Yes [If so, please list here.:] □ No □ Don't know

◈ **Has your pet had any operations?** ([23])
□ Yes [If so, please list here.:] □ No □ Don't know

◈ **What's your pet's problem today?** ([24])

Appendix for Unit 2: Combination Vaccines for dogs and cats

犬の混合ワクチンの種類

	2種	3種	4種	5種	6種	7種	8種	9種	10種	11種
canine distemper	○	○	○	○	○	○	○	○	○	○
canine hepatitis		○	○	○	○	○	○	○	○	○
canine adenovirus type 2		○	○	○	○	○	○	○	○	○
canine parvovirus	○			○	○	○	○	○	○	○
canine parainfluenza			○	○	○	○	○	○	○	○
canine coronavirus					○		○	○	○	○
canine leptospirosis leptospira icterohaemorrhagiae						○	○	○	○	○
leptospira canicola						○	○	○	○	○
leptospira hebdomadis								○	○	○
leptospira autumnalis									○	○
leptospira australis										○

＊この表は一例に過ぎず、異なるレプトスピラの血清型を組み合わせたワクチンもあります。

[A] レプトスピラ症の血清型の発音 

レプトスパァイラ インテェロガンズ スィアラヴァ イクティェロヘェ モリィジァィ
イクテロヘモラジー／黄疸出血型：leptospira (interrogans serovar) icterohaemorrhagiae

レプトスパァイラ インテェロガンズ スィアラヴァ キィャニコラ
カニコーラ型：leptospira (interrogans serovar) canicola

レプトスパァイラ インテェロガンズ スィアラヴァ ヘェブドゥマディス
ヘブドマディス型：leptospira (interrogans serovar) hebdomadis

レプトスパァイラ インテェロガンズ スィアラヴァ オゥタァ ム ナ リス
オータムナリス型：leptospira (interrogans serovar) autumnalis

レプトスパァイラ インテェロガンズ スィアラヴァ アゥストラァ リス
オーストラリス型：leptospira (interrogans serovar) australis

猫の混合ワクチンの種類

	3種	4種	5種	7種
feline viral rhinotracheitis (FVR) feline herpesvirus 1 (FeHV-1)	○	○	○	○
feline calicivirus (FC-7)	○	○	○	○
feline panleukopenia (FPL)	○	○	○	○
feline leukemia virus infection (FeLV)		○	○	○
chlamydia felis			○	○
feline calicivirus (FC-28)				○
feline calicivirus (FC-64)				○

＊この表は一例に過ぎません。

[B] Unit 2 未収語の発音 

猫白血病ウイルス感染症：feline（フィーライン） leukemia（ルゥーキィーミィア） virus（ヴァィァラス） infection（インフェクシャン）

猫クラミジア感染症：chlamydia（クラァミィディア） felis（フェリス）

単独接種のワクチン

猫免疫不全ウイルス感染症：（猫エイズ） feline（フィーライン） immunodeficiency（イミュゥノォゥディフィシャンシィ） virus（ヴァィァラス） [infection]（インフェクシャン）
(FIV)

Appendix for Unit 6: Names of Diseases and Symptoms

1. Does he/she have [　　] ? 「…の症状はありますか？ / …を患っていますか？」

ダァイアリ ィーア　アラルヂィズ　ダァイアビィーティーズ　スタァマァケ ェイク
diarrhea（下痢） / allergies（アレルギー） / diabetes（糖尿病） / a stomachache（腹痛） /

インフェクシャン　サァイアロイドゥ プラァブラァムズ　ディメ ェンシィア
an ear infection（耳の感染症） / thyroid problems（甲状腺の異常） / dementia（認知症） /

ブレェ ス　アズ マァ　フィーヴァ　コォーフ　ラァシュ
bad breath（口臭） / asthma（喘息） / a fever（熱） / a cough（咳） / a rash（発疹） /

スキン レェドゥニィス　ラァニィ ノォウズ
any skin redness（皮膚の発赤） / a runny nose（鼻水）

2. Is he/she [　　] ? 「…していますか」

ヴァミティング　リィンピング　コォーフィング
vomiting（嘔吐している） / limping（足を引きずっている） / coughing（咳をしている） /

カァンスタァペェイティットゥ　スニィージィング　ペェイン
constipated（便秘している） / sneezing（くしゃみしている） / in pain（痛がっている） /

フウィージィング　ブレェス
wheezing（ゼーゼーしている） / short of breath（息切れしている）

3. How is his/her [　　] ? 「…はどんな具合ですか？」

ビィヘェイヴィア　アパァタァイトゥ　ストゥール　アクティヴィティレェヴァル
behavior（行動） / appetite（食欲） / stool（便） / activity level（活動レベル） /

エナァヂィ レェヴァル　ウェイトゥ　ユゥアリィン　ムゥードゥ
energy level（エネルギーレベル） / weight（体重） / urine（尿） / mood（機嫌）

4. Is/Are his/her (right/left) [　　] swollen (and painful)?
「…は腫れて（痛そうにして）いますか？」

eye(s) / face / leg(s) / stomach / ear(s) / nose / knee(s) / pad(s) / gum(s) / paw(s)

❖ 獣医師による診断結果を知らせる場合は、次のような表現を使います。さまざまな病名と共に覚えておきましょう。

ダァイァグノォウズドゥ
The vet diagnosed him/her with ＿＿＿. 「獣医の診断によると…ということです」

ギァストゥロゥエンタラァイティス　ニュゥモォウニャ　カァンヂァンクタァヴァイティス　キァンサァ
gastroenteritis（胃腸炎）/ pneumonia（肺炎）/ conjunctivitis（結膜炎）/ cancer（癌）/

ハァートゥワァームズ　ハァートゥ フェイリャア　ボォウン フラァクチャ　スプレェイン
heartworms（フィラリア）/ heart failure（心不全）/ bone fracture（骨折）/ sprain（捻挫）/

リィンフォウマ　アースラァイティス　パァラァシィティク ダァーマァタァイティス
lymphoma（リンパ腫）/ arthritis（関節炎）/ parasitic dermatitis（寄生虫性皮膚炎）/

エパレェプシィ　ペェリィアダァントゥル ディズィーズ　ケェンル コォーフ
epilepsy（てんかん）/ periodontal disease（歯周病）/ kennel cough（ケネルコフ）/

アダァスンズ ディズィーズ　キァタラァクトゥ
food allergies（食物アレルギー）/ Addison’s disease（アジソン病）/ cataract（白内障）/

グロォーコォウマ　ケェラァタァイティス　リィングワァーム
glaucoma（緑内障）/ keratitis（角膜炎）/ dry eye（ドライアイ）/ ringworm（白癬）/

メェインヂ　ダァーマタァファイトォウシィス　インテェスタァンル ブラァキィヂュ
mange（疥癬）/ dermatophytosis（皮膚糸状菌症）/ intestinal blockage（腸閉塞）/

パァイァダァーマ　セェバァリィーア　ギァストゥリィクトォーシャン
pyoderma（膿皮症）/ seborrhea（脂漏症）/ gastric torsion（胃捻転）/

スタァマァク アルサァ　シィスタァイティス　ブラァダァ ストォウン
stomach ulcer（胃潰瘍）/ cystitis（膀胱炎）/ bladder stone（膀胱結石）/

オゥタァイティス インタァーナ　オゥタァイティス ミィーディア　オゥタァイティス エクスタァーナ
otitis interna（内耳炎）/ otitis media（中耳炎）/ otitis externa（外耳炎）

Appendix for Unit 9-1: Euthanasia Consent Form

"Euthanasia Consent Form (安楽死の同意書)" の下線部（1）～ (6)、(8) に枠内から適切な語句を選んで書き入れなさい。(7) については、与えられた語を並べ替えて正しい文章を完成なさい。

Euthanasia Consent Form

Today's Date ....

Owner's Name: ....
Address: ....
Phone: (home) .... (mobile) ....
Pet's Name: ....
Breed: .... Color: .... Weight: ....
Species: .... Sex: .... Age: ....

I, the undersigned, (1) ________________（証明する）that I am the owner (or authorized agent for owner) of the animal (2) ________________（上記の）, request and consent euthanasia to be (3) ________________（行われる）on said animal.

I hereby give Dr. .... and his/her assistants or staff full and complete (4) ________________（権限）to euthanize said animal in a (5) ________________（人道的方法）. Furthermore, I forever release the veterinarian, his/her assistants or staff from any and all (6) ________________（法的責任）of said euthanasia.

(7) [fully / ending / I / of / the life / euthanasia / that / the act / an animal / understand / is / of]（私は安楽死が動物の命を終わらすす行為であることを重々承知しております）

____

in a painless way to prevent any (8) ________________________（不要な苦しみ）.
To the best of my knowledge, the information provided is accurate and complete.

Signature: .... Date: ....

liability / authority / certify / unnecessary suffering / described above / performed / humane manner

Appendix for Unit 9-2: Mini Conversation（動物葬祭ディレクターに相談する）

[A] 動物葬祭ディレクターと愛猫を亡くしたオーナーさんの会話を練習しなさい。 

Ms. Field and her family returned to the clinic later that evening and Henry was put to sleep quietly with his family around him. Now Ayaka is helping them through the next steps. Ayaka gave them some pamphlets to look for local pet crematoriums and the Field family took Henry home. The next day Ms. Field telephones the Pets' Rainbow Crematorium.

Receptionist : Hello, this is the Pets' Rainbow Crematorium. How may I help you?

Ms. Field : Hello, My name is [1]Field, Mary Field. My cat passed away last night and I'd like to ask about having him cremated.

Receptionist : I'm very sorry to hear that, [1]Ms. Field. You have my deepest sympathies.

Ms. Field : Thank you. We'd just like a very simple cremation. Could you tell me a little about it?

Receptionist : Do you have our pamphlet? There is a simple cremation described on page 10.

Ms. Field : It looks just right. Can you come tomorrow?

Receptionist : Certainly. I'll need to take down a few details first.
Your full name is [1]Mary Field, right?

Ms. Field : Yes.

Receptionist : And the address where we may collect him?

Ms. Field : He's at home. [2]That's 1-2-3, Fujimi-cho, Koto-ku, Tokyo.

Receptionist : Could you tell me your telephone number, please?

Ms. Field : It's [3]080-1234-56XX. When will you be coming?

Receptionist : If it's convenient for you, we can be there at [4]10:00 o'clock tomorrow morning.

Ms. Field : That's perfect. Can we come with him?

Receptionist : Yes, please do. You can ride in the hearse or you can follow in your own car.

Ms. Field : Fine. See you [4]tomorrow at 10:00, then.

Receptionist : Yes. Please bring your name stamp if you have one as there is some paperwork to be filled in. Goodbye for now.

Ms. Field : Goodbye.

[B] ペアになり、1～3の下線部はそれぞれ自分の名前、住所、電話番号に変え、4の下線部は日にちや時間を変えて、もう一度会話を練習しなさい。

Appendix for Unit 10: Names of Breeds: Cats

枠内のヒントを参考にして 1 ～ 12 のアルファベットの文字を並べ替えて猫種名を完成させ、その猫種の写真を下から選び（　　）内に記号で答えなさい。

Exotic Shorthair / Norwegian Forest Cat / Russian Blue /
Munchkin / Scottish Fold / Siamese / Abyssinian / Persian /
American Shorthair / Maine Coon / Ragdoll / Bengal

1. unikhMnc ______________________ (　　)
2. tFSilsh odcot ______________________ (　　)
3. iMaoonenc ______________________ (　　)
4. mioAear itchnrrSah ______________________ (　　)
5. ianPrse ______________________ (　　)
6. gBnale ______________________ (　　)
7. alnRss uueiB ______________________ (　　)
8. ldagRlo ______________________ (　　)
9. ShroE hirxcatoit ______________________ (　　)
10. Freeot sgintrC oNawa ______________________ (　　)
11. sSiaeme ______________________ (　　)
12. yAnabniiss ______________________ (　　)

a.

b.

c.

d.

e.

f.

g.

h.

i.

j.

k.

l.

Parts of the Body (The Dog)

番号に相当する犬の体の部位の名称を、それぞれ英語と日本語の語群の中から選び、下線部に英語を、（　）内に日本語を書き入れなさい。

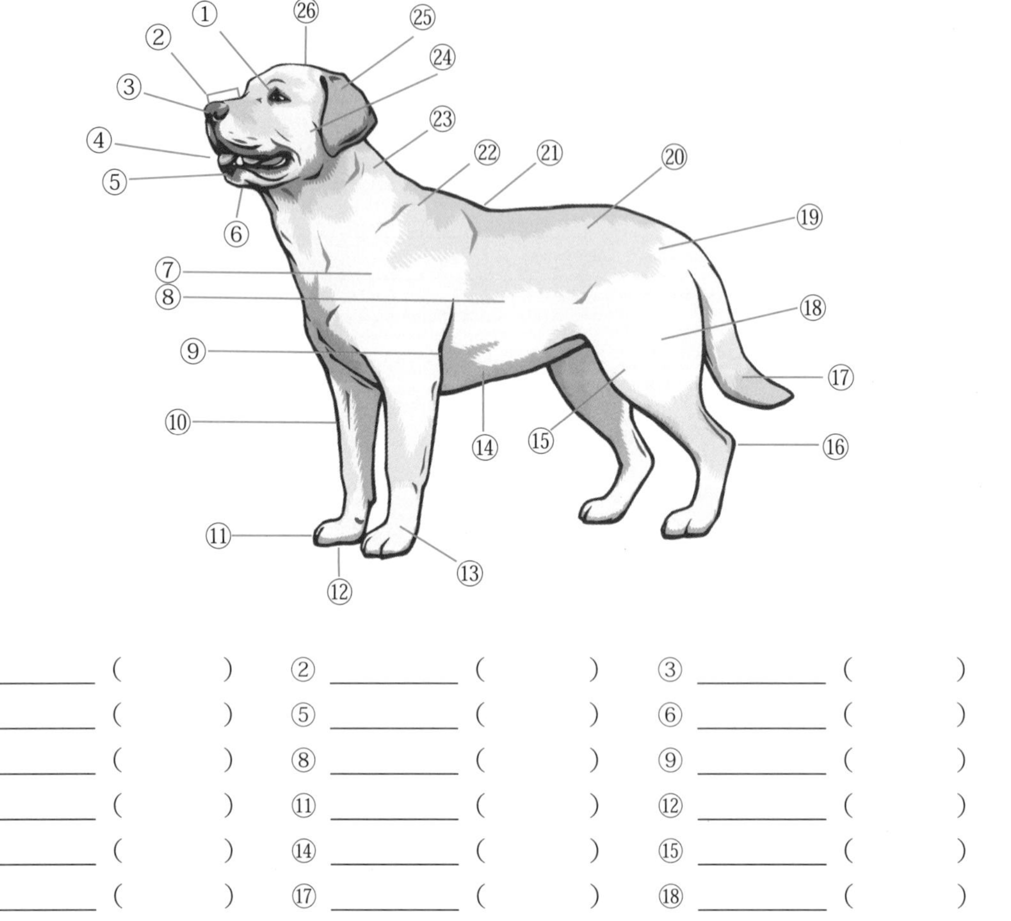

① ＿＿＿＿＿（　　　）　② ＿＿＿＿＿（　　　）　③ ＿＿＿＿＿（　　　）

④ ＿＿＿＿＿（　　　）　⑤ ＿＿＿＿＿（　　　）　⑥ ＿＿＿＿＿（　　　）

⑦ ＿＿＿＿＿（　　　）　⑧ ＿＿＿＿＿（　　　）　⑨ ＿＿＿＿＿（　　　）

⑩ ＿＿＿＿＿（　　　）　⑪ ＿＿＿＿＿（　　　）　⑫ ＿＿＿＿＿（　　　）

⑬ ＿＿＿＿＿（　　　）　⑭ ＿＿＿＿＿（　　　）　⑮ ＿＿＿＿＿（　　　）

⑯ ＿＿＿＿＿（　　　）　⑰ ＿＿＿＿＿（　　　）　⑱ ＿＿＿＿＿（　　　）

⑲ ＿＿＿＿＿（　　　）　⑳ ＿＿＿＿＿（　　　）　㉑ ＿＿＿＿＿（　　　）

㉒ ＿＿＿＿＿（　　　）　㉓ ＿＿＿＿＿（　　　）　㉔ ＿＿＿＿＿（　　　）

㉕ ＿＿＿＿＿（　　　）　㉖ ＿＿＿＿＿（　　　）

English names

abdomen (or belly) / tail / knee (or stifle) / shoulder / head / mouth / back / paw / chest (or brisket) / muzzle / neck / pads / nose / thigh / eye / loin / ear / forearm / lip (or flew) / hock / cheek / pastern / withers / elbow / rump / jaw

Japanese names

腰 / 前腕 / 口吻 / 尾 / 鼻 / 腹部 / 口唇 / 肉球 / 頬 / 膝 / 顎 / 首 / 大腿
足 / 臀部 / 胸部 / 背 / 目 / 飛節 / 頭部 / キ甲 / 肩 / 耳 / 肘 / 口 / 中手

Parts of the Body (The Internal Organs of Dogs)

番号に相当する臓器の名称を、それぞれ英語と日本語の語群の中から選び、下線部に英語を、（　）内に日本語を書き入れなさい。

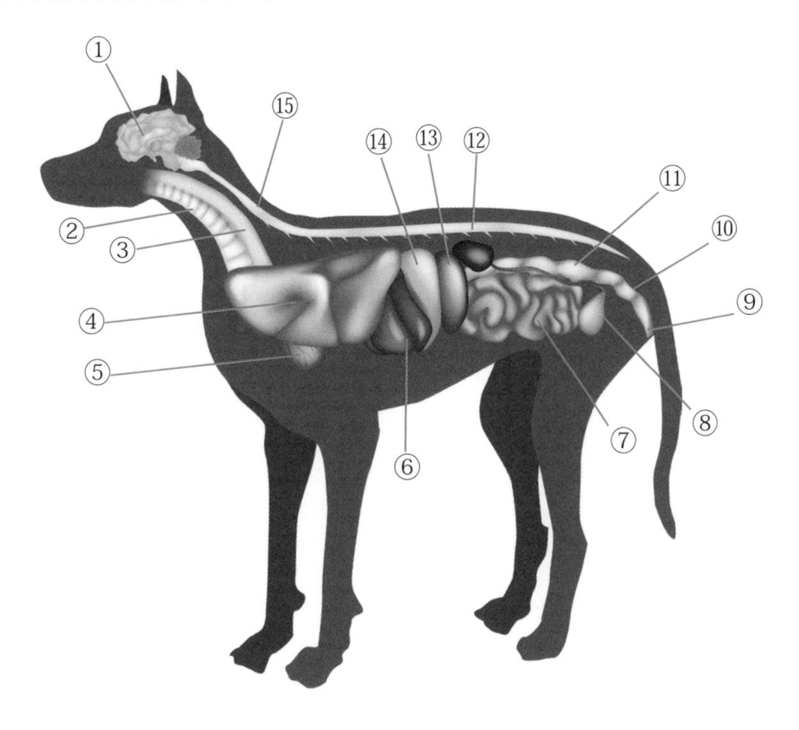

① ＿＿＿＿＿ （　　　）	② ＿＿＿＿＿ （　　　）	③ ＿＿＿＿＿ （　　　）
④ ＿＿＿＿＿ （　　　）	⑤ ＿＿＿＿＿ （　　　）	⑥ ＿＿＿＿＿ （　　　）
⑦ ＿＿＿＿＿ （　　　）	⑧ ＿＿＿＿＿ （　　　）	⑨ ＿＿＿＿＿ （　　　）
⑩ ＿＿＿＿＿ （　　　）	⑪ ＿＿＿＿＿ （　　　）	⑫ ＿＿＿＿＿ （　　　）
⑬ ＿＿＿＿＿ （　　　）	⑭ ＿＿＿＿＿ （　　　）	⑮ ＿＿＿＿＿ （　　　）

English names

heart / rectum / lung / small intestine / spleen / trachea / liver / anus / esophagus / kidney / spinal cord / colon / brain / stomach / bladder

Japanese names

小腸 / 肺 / 脾臓 / 心臓 / 脊髄 / 膀胱 / 胃 / 腎臓 / 肝臓 / 脳 / 肛門 / 食道 / 直腸 / 気管 / 結腸

NOTE

NOTE

NOTE

Animal Companions [B-920]
動物専門職のための総合英語

1刷	2021年3月12日
著　者	Susan Williams　スーザン・ウィリアムズ Midori Asai　浅井　みどり
発行者	南雲　一範　Kazunori Nagumo
発行所	株式会社　南雲堂 〒162-0801　東京都新宿区山吹町361 NAN'UN-DO Co., Ltd. 361 Yamabuki-cho, Shinjuku-ku, Tokyo 162-0801, Japan 振替口座 : 00160-0-46863 TEL:　03-3268-2311（営業部：学校関係） 03-3268-2384（営業部：書店関係） 03-3268-2387（編集部） FAX:　03-3269-2486
編集者	加藤　敦
組　版	中西　史子
装　丁	浜野　史
イラスト	浜野　史（p. 4-5, 9, 27, 37, 39, 49, 52, 64, 69, 77, 84, 90-92）
検　印	省　　略
コード	ISBN978-4-523-17920-7　C0082

Printed in Japan

E-mail : nanundo@post.email.ne.jp
URL : http://www.nanun-do.co.jp/